Michael Bodensteiner

Lewis Acid/Base-Stabilized Phosphanylalanes

Michael Bodensteiner

Lewis Acid/Base-Stabilized Phosphanylalanes

and Crystal Structure Determinations

Südwestdeutscher Verlag für Hochschulschriften

Impressum / Imprint

Bibliografische Information der Deutschen Nationalbibliothek: Die Deutsche Nationalbibliothek verzeichnet diese Publikation in der Deutschen Nationalbibliografie; detaillierte bibliografische Daten sind im Internet über http://dnb.d-nb.de abrufbar.

Alle in diesem Buch genannten Marken und Produktnamen unterliegen warenzeichen-, marken- oder patentrechtlichem Schutz bzw. sind Warenzeichen oder eingetragene Warenzeichen der jeweiligen Inhaber. Die Wiedergabe von Marken, Produktnamen, Gebrauchsnamen, Handelsnamen, Warenbezeichnungen u.s.w. in diesem Werk berechtigt auch ohne besondere Kennzeichnung nicht zu der Annahme, dass solche Namen im Sinne der Warenzeichen- und Markenschutzgesetzgebung als frei zu betrachten wären und daher von jedermann benutzt werden dürften.

Bibliographic information published by the Deutsche Nationalbibliothek: The Deutsche Nationalbibliothek lists this publication in the Deutsche Nationalbibliografie; detailed bibliographic data are available in the Internet at http://dnb.d-nb.de.

Any brand names and product names mentioned in this book are subject to trademark, brand or patent protection and are trademarks or registered trademarks of their respective holders. The use of brand names, product names, common names, trade names, product descriptions etc. even without a particular marking in this works is in no way to be construed to mean that such names may be regarded as unrestricted in respect of trademark and brand protection legislation and could thus be used by anyone.

Coverbild / Cover image: www.ingimage.com

Verlag / Publisher:
Südwestdeutscher Verlag für Hochschulschriften
ist ein Imprint der / is a trademark of
AV Akademikerverlag GmbH & Co. KG
Heinrich-Böcking-Str. 6-8, 66121 Saarbrücken, Deutschland / Germany
Email: info@svh-verlag.de

Herstellung: siehe letzte Seite /
Printed at: see last page
ISBN: 978-3-8381-3411-6

Zugl. / Approved by: Regensburg, Uni, Diss., 2011

Copyright © 2012 AV Akademikerverlag GmbH & Co. KG
Alle Rechte vorbehalten. / All rights reserved. Saarbrücken 2012

Meinen Eltern gewidmet

The important thing in science is not so much to obtain new facts as to discover new ways of thinking about them.

Sir William Lawrence Bragg (1890 – 1971)

Contents

1. Introduction 1
1.1. Group 13/15-Compounds 1
1.1.1. Hydrogen Storage Applications and other new Developments 1
1.1.2. The Concept of Lewis Acid/Base-Stabilization 3
1.1.3. Oligomers of Hydrogen Elimination Reactions 5
1.2. Crystallography 8
1.2.1. Electron and Neutron Diffraction 8
1.2.2. Single Crystal X-ray Diffraction 9
2. Research Objectives 10
3. Synthetic Section 11
3.1. Isomerization of LA/LB-stabilized Phosphanylalanes 11
3.2. The Controlled Oligomerization of [{(CO)$_5$W}H$_2$PAlH$_2$·NMe$_3$] (**4**) 18
3.3. A new Cage Motif of Phosphanylalanes employing the Lewis Base NMe$_2$Et 29
3.4. Lewis acid-free Phosphanylalanes 32
3.5. Introduction of a *N*-heterocyclic Carbene as Lewis Base 37
3.6. Stiba- and Bismaboranes 39
4. Crystallographic Section 41
4.1. General Procedures 41
4.1.1. Sample Handling 41
4.1.2. Data Collection 41
4.1.3. Data Processing 42
4.1.4. Space Group Determination 42
4.1.5. Structure Solution and Refinement 43
4.2. Disorder in Mixed Crystals 45
4.3. Twinning 48
4.3.1. Pseudo-merohedral Twinning 49

4.3.2.	Merohedral in Combination with pseudo-merohedral Twinning	50
4.4.	Space Group Problem	51
4.5.	Disordered Solvent Treatment Applying SQUEEZE	55
4.6.	Modulated Structure	57
5.	Experimental Section	60
5.1.	General Methods	60
5.2.	Alternative Synthesis of [{(CO)$_5$W}H$_2$PAlH$_2$·NMe$_3$] (**4**)	60
5.3.	[{(CO)$_5$W}HPAlH·NMe$_3$]$_3$ (**5**)	61
5.4.	[⟨{(CO)$_5$W}HPAlH·NMe$_3$⟩$_2$⟨(CO)$_5$WPAl·NMe$_3$⟩] (**6**)	63
5.5.	Alternative Synthesis of [{(CO)$_5$WPH$_2$}(Me$_3$N)AlPH{W(CO)$_5$}]$_2$ (**7**)	65
5.6.	[{(CO)$_5$WPH$_2$}(Me$_2$EtN)AlPH{W(CO)$_5$}]$_2$ (**10**) and [⟨{W(CO)$_5$}HPAl(Me$_2$EtN)⟩$_2$ μ-⟨{(CO)$_5$WPH}$_2$Al(Me$_2$EtN)⟩] (**11**)	65
5.7.	Synthesis of (Me$_3$Si)$_2$PAlH$_2$·NMe$_3$ (**12**)	68
5.8.	Synthesis of [(Me$_3$Si)PAlH·NMe$_3$]$_2$ (**13**)	70
5.9.	Synthesis of H$_3$Al·NHC^{Me} (**14**)	70
5.10.	Synthesis of (Me$_3$Si)$_2$SbBH$_2$·NMe$_3$ (**15**)	72
6.	Summary and Conclusions	73
6.1.	Synthetic Results	73
6.2.	Crystallographic Results	76
7.	Appendix	78
7.1.	Supporting Data	78
7.2.	List of Compounds	78
7.3.	List of Abbreviations	79
8.	Literature	81

1. Introduction

1.1. Group 13/15-Compounds

Compounds with a direct bond between an element of group 13 (E' = boron, aluminium, gallium, indium) and an element of group 15 (E = nitrogen, phosphorus, arsenic, antimony) are promising materials in inorganic research. They are isoelectronic to group 14 elements, and show a large variety of technical applications replacing those or widening their application range. For instance both structural modifications of boron nitride can be used in place of their carbon analogues diamond, as an abrasive, and graphite as a lubricant.[1] Moreover boron nitride can also form nanomeshes and nanotubes like carbon.[2,3] Furthermore, the semiconducting properties allow them to substitute silicon and germanium in lasers, solar panels, light emitting (LED) and photo diodes.[4] The technical importance of lasers is mirrored by the fact the 2000 Nobel Prize in Physics was awarded for the research on binary and ternary layers of Al/Ga and P/As.[5,6] Those lasers offer a large wavelength variety depending on their chemical composition. Furthermore, it is a long known fact that solar cells consisting of any bulk material cannot exceed 31% efficiency.[7] New efforts employ GaAs/InAs quantum dots to bypass this limit.[8] In everyday life the white LED is probably the most prominent application of 13/15-compounds.[9] It is a standard component of torches and displays, which often consist of GaN or InGaN coated with a layer of a phosphorescent material. Their biggest advantage is the efficiency (70 lm/W) compared to standard light bulbs (12 lm/W). Single die devices have been shown to be able to produce more than 100 lm.

1.1.1. Hydrogen Storage Applications and other new Developments

The current main focus in 13/15-chemistry is hydrogen storage mainly based on $H_3B \cdot NH_3$, since weight is important for possible applications, like car reservoirs.[10,11] The scope of research to liberate hydrogen from this system includes acids,[12] transition metal catalysts,[13-16] nanoparticles[17,18] as well as ionic liquids[19] (Equation (1)).

$$H_3B \leftarrow NH_3 \xrightarrow{[cat]} H_{3-x}BNH_{3-x} + xH_2 \quad x = 1-3 \quad (1)$$

The major problem in these systems is the irreversibility of this reaction. If all three equivalents of hydrogen are eliminated the final product is boron nitride, the chemically inert nature of which inhibits the re-addition of H_2. However, compounds containing a phosphorus-boron double bond like $^tBu_2P=B(C_6F_5)_2$ also show hydrogen activation properties.[20] If an interaction between the group 13 and group 15 element is inhibited by sterically demanding substituents such compounds are generally referred to as 'frustrated Lewis pairs' (FLPs).[21,22] The combination of a phosphane with a borane in FLPs often leads to a heterolytic cleavage of dihydrogen. In FLPs, the Lewis acid (LA) and Lewis base (LB) functions can be separated in different molecules, or both functions can be connected within the same molecule (Equations (2) and (3)).

$$R_3P + BR'_3 \xrightarrow{+ H_2} [R_3PH]^+[HBR'_3]^- \qquad \begin{array}{l} R = {}^tBu, 2,4,6\text{-}C_6H_2Me_3 \\ R' = C_6F_5, C_6H_5 \end{array} \qquad (2)$$

(3)

In the case of the latter systems the hydrogen addition is reversible (Equation (3)), but the amount of stored hydrogen is only 0.25 mass per cent. It has been shown that those FLPs are not limited to reactions with hydrogen, but can also activate and reversibly store other small molecules, e.g. CO_2 (Equations (4) and (5)).[22]

(4)

(5)

1. INTRODUCTION

Even non-frustrated phosphorus-boron systems undergo hydrogen elimination forming oligomers and polymers employing a Rh(I) catalyst.[23,24] It has been shown by *Adolf* in our group, that such a polymer can be depolymerized by addition of a Lewis base[25] (Equation (6)).

$$PhPH_2 \cdot BH_3 \xrightarrow[-H_2]{[Rh(I)]} [PhPH-BH_2]_n \xrightarrow{LB} PhPH-BH_2(LB) \quad (6)$$

LB = 1,3-dimethylimidazol-2-ylidene or 4-(dimethylamino)pyridine

However, not only the B/N and B/P systems show interesting reactivities. Among the higher homologues, aluminium phosphorus systems in particular have been found to activate C≡C triple bonds (Equation (7)).[26] Furthermore, the strong affinity of aluminium towards oxygen is used to bind CO_2 (Equation (8)).[27]

$$Aryl-P(C \equiv C-CMe_3)_2 + HAlR_2 \longrightarrow \text{[cyclic product]} \quad (7)$$

Aryl = C_6H_5, 2,4,6-$C_6H_2Me_3$
R = CMe_3, CH_2CMe_3

$$tmp_2AlP(SiMe_3)_2 \xrightarrow{CO_2} tmp_2Al\ominus\text{-O-O-}\oplus C-P(SiMe_3)_2 \quad (8)$$

tmp = 2,2,6,6-tetramethylpiperidine

1.1.2. The Concept of Lewis Acid/Base-Stabilization

Besides H_2 storage purposes, 13/15-compounds carrying hydrogen substituents are only rarely found in this chemistry. Usually bulky groups have to be used to avoid head-to-tail polymerization. This is caused by a lone-pair at the group 15 element together with a free p-orbital at the group 13 element.

Employing sterically demanding substituents, immediate intermolecular polymerization can be inhibited. Applying this technique allows monomeric 13/15-compounds to be obtained (Figure 1).

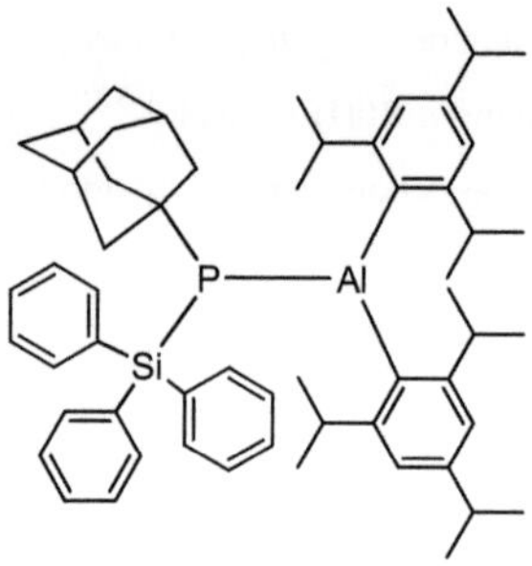

Figure 1: Example for a stable, sterically hindered monomeric phosphanylalane.[28]

For reason of the beforehand mentioned instability, the exclusively hydrogen-substituted parent compounds H$_2$E–E'H$_2$ (**A**) have only been studied theoretically (Scheme 1).[29-33] However, a stabilization of those can be achieved by blocking the acceptor and donor functions at the group 13 and 15 element using Lewis bases and Lewis acids (**D**). By this method, developed in our research group, polymerization can also be avoided (Equation (9)).

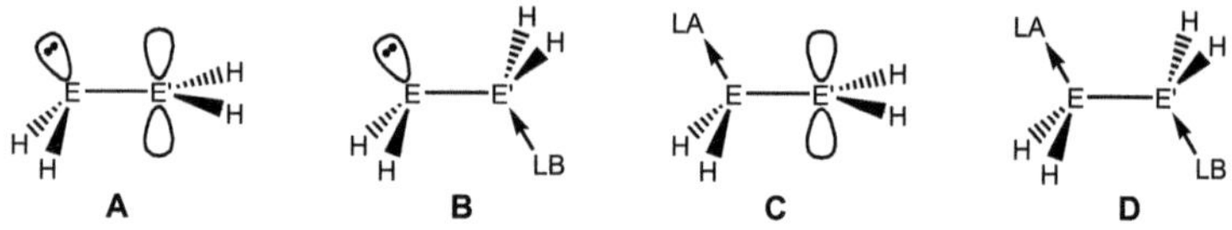

Scheme 1: Different types of hydrogen-substituted pentelyltrielanes.

Employing this concept, the first stabilized phosphanyl- and arsanylboranes[34] as well as phosphanylalanes and -gallanes[35] could be synthesized, using M(CO)$_5$ (M = Cr, W) or E'(C$_6$F$_5$)$_3$ (E' = B, Ga) as LA, and amines or an *N*-heterocyclic carbene as LB, respectively. For the synthesis of those boron compounds, salt elimination reactions are employed to obtain the desired products.

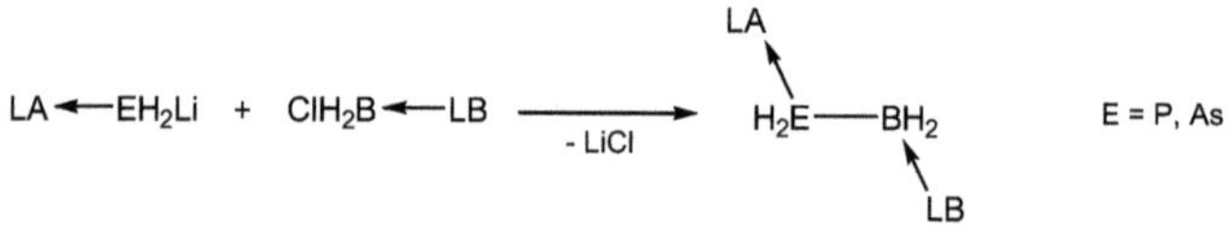

$$\text{LA} \leftarrow \text{EH}_2\text{Li} + \text{ClH}_2\text{B} \leftarrow \text{LB} \xrightarrow{-\text{LiCl}} \text{LA} \rightarrow \text{H}_2\text{E} - \text{BH}_2 \leftarrow \text{LB} \quad \text{E = P, As} \quad (9)$$

In the case of the higher group 13 homologues, hydrogen eliminations lead to the LA/LB-stabilized 13/15-compounds (Equation (10)). Aluminium and gallium are the most electropositive elements in group 13 and phosphorus is quite electronegative, hence, this relatively large difference enforces the hydrogen elimination reactions between the hydridic

1. INTRODUCTION

and the protic hydrogens at the aluminium and phosphorus, respectively. The electronegativities of boron and phosphorus are quite similar, thus, no comparable hydrogen elimination occurs, due to the low polarization of the different hydrogen atoms.

$$LA \leftarrow PH_3 + H_3E' \leftarrow LB \xrightarrow{-H_2} \begin{array}{c} LA \\ | \\ H_2P-E'H_2 \\ | \\ LB \end{array} \quad E' = Al, Ga \quad (10)$$

Computations considered only LB- (**B**) or LA-stabilized (**C**) derivatives to be stable. Experimentally the synthesis of the first LB-only-stabilized, hydrogen substituted pentelylboranes $H_2EBH_2 \cdot NMe_3$ (E = P, As) could prove this theory (Equation (11)).[36-38]

$$(Me_3Si)_2ELi + ClH_2B \leftarrow NMe_3 \xrightarrow{-LiCl} \begin{array}{c} Me_3Si \\ \diagdown \\ E-BH_2 \\ \diagup \\ Me_3Si \quad NMe_3 \end{array} \xrightarrow[-Me_3SiOMe]{+MeOH} \begin{array}{c} H \\ \diagdown \\ E-BH_2 \\ \diagup \\ H \quad NMe_3 \end{array} \quad (11)$$

E = P, As

All efforts to synthesize LA-only-stabilized compounds of type **C** failed to date. Similar monomers of the higher group 13 homologues cannot be synthesized due to their strong tendency to eliminate hydrogen under oligomerization.

1.1.3. Oligomers of Hydrogen Elimination Reactions

Oligomers and polymers of group 13 and group 15 elements show versatile structural motifs, which have widely been studied both experimentally and theoretically.[39-42] However, the mechanisms of the reactions that result in such oligomers have only rarely been investigated.

Theoretically, the non-existent hydrogen-only substituted parent compound **A** could undergo a dimerization forming a four-membered ring motif, which again loses hydrogen forming a cube shaped structural motif (Scheme 2). Such a heterocubane motif has already been reported for the phosphanylalane [^{i}BuAlP(SiPh$_3$)]$_4$.[43] In the case of the fully hydrogen substituted cubane a final hydrogen elimination step would then lead to the binary 13/15-material. In the case of the LA/LB-stabilized compound the cubane motif is already the hydrogen free derivative and hence the last step of the possible successive hydrogen elimination mechanism.

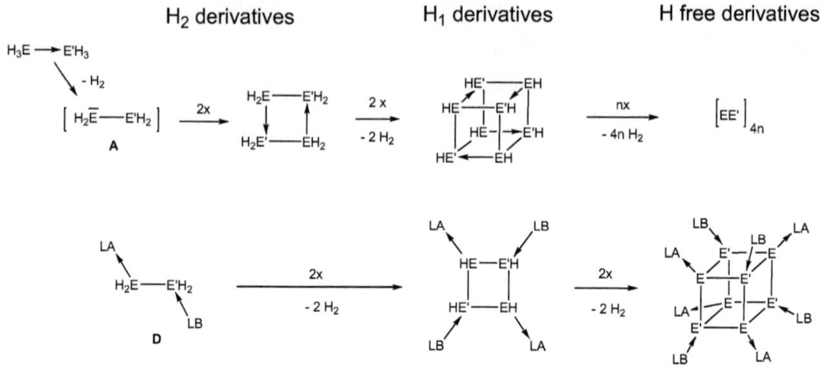

Scheme 2: Schematic comparison of the dimerization and subsequent processes starting from unprotected and from Lewis acid/base-stabilized 13/15-compounds. E = group 15 element, E' = group 13 element.

Another plausible reaction pathway is the trimerization (Scheme 3). Hereby the parent compound **A** reacts towards a six-membered heterocycle instead of a four-membered one. In this case two of these trimers could combine forming a hexagon structural motif. Examples for this has been described by *Hänisch* for the chlorine- and silyl-substituted compounds [ClAlPR]$_6$ (R = Si^{i}Pr$_3$, Si^{i}Pr$_2$Me).[44]

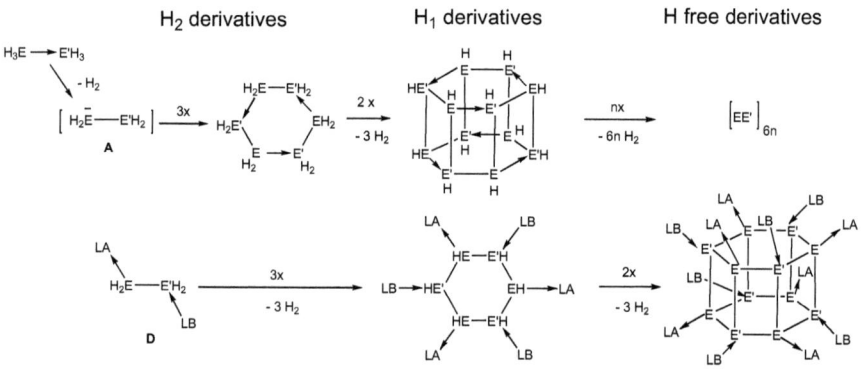

Scheme 3: Schematic comparison of the trimerization and subsequent hexamerization processes starting from unprotected and from Lewis acid/base-stabilized 13/15-compounds. E = group 15 element, E' = group 13 element.

1. INTRODUCTION

In our group, *Vogel* already postulated several different oligomerization products from NMR studies on crude reaction mixtures during the formation of the first LA/LB-stabilized compounds for the higher homologues phosphanylalane and -gallane.[35] For the aluminium derivative he was able to characterize a four-membered ring compound and could also obtain a product containing a six-membered ring motif characterized by a low-quality X-ray experiment (Equation (12)).[45]

$$(OC)_5W \leftarrow PH_3 + H_3Al \leftarrow NMe_3 \xrightarrow{-H_2} [\text{four-membered ring}] + [\text{six-membered ring}] \quad (12)$$

A new structural motif for phosphanylalanes was found by *Schwan* in our group changing the Lewis base to *N,N*-dimethylaminopyridine (Equation (13)).[38]

$$(OC)_5W \leftarrow PH_3 + H_3Al \leftarrow \text{dmap} \xrightarrow{-H_2} [\text{six-membered ring with dmap}] \quad (13)$$

dmap = 4-dimethylaminopyridine

During my diploma thesis research, new four-membered rings could be synthesized by changing the LB to triethylamine and by replacing one hydrogen substituent at the phosphane by a phenyl group (Equation (14)).[46]

$$(OC)_5W \leftarrow PRH_2 + H_3Al \leftarrow NR'_3 \xrightarrow{-H_2} [\text{four-membered ring}] \quad (14)$$

R = H, Ph; R' = Me, Et

The hydrogen elimination processes often depend on the polarity of the solvent and the temperature. Shape, size, Lewis acidity and -basicity, respectively, can also influence the process of oligomerization.

1.2. Crystallography[47]

Scattering techniques have become very important for many fields of chemistry. They are applied in different experiments to determine composition, purity, structure, and other properties of compounds and materials. For example, small and wide angle X-ray scattering is used to investigate materials like polymers or colloidal solutions on scales ranging from a few Ångströms to one micrometer.[48] Powder diffraction techniques are employed to identify or characterize substances and to obtain the purity of polycrystalline samples.[49] In rare cases it is even possible to determine the structure by applying a method developed by *Rietveld*.[50] However, the most common techniques for structure determination purposes employ electron, neutron or X-ray beams at single crystals. Compared to other analytical techniques, these methods are not limited to certain elements and can be applied to any crystalline sample.

1.2.1. Electron and Neutron Diffraction

An electron beam interacts both with the nuclei and the electron shells of the atoms, leading to a strong absorption of the beam. Hence, electron diffraction can only be applied to determine the structures of small molecules in gas phase or very small or thin crystalline samples. This can be an advantage, allowing a structure determination for samples that are too small for other diffraction techniqes. However, other problems of this method are thermal impact due to absorption, lack of data for geometry reasons and the mathematical description of the elastic and inelastic scattering at both the electron shell and the nuclei. These facts lead to problems in structure solution and refinement.

Neutron diffraction requires a neutron source, mostly a nuclear reactor. The neutrons are only diffracted at the nuclei, and the obtained intensity is independent of the diffraction angle (in comparison to electron and X-ray diffraction). Thus, one advantage compared to the other methods is the exact determination of hydrogen positions. Furthermore, neighbouring elements in the periodic table can be distinguished, what is sometimes a problem in X-ray crystallography (see below). It even allows different isotopes to be identified. The disadvantages are that relativly large sample amounts (single crystals on millimeter- and powders on gramm-scales) are required and long time experiments have to be carried out (usually days). In addition, another problem is the availability of neutron sources.

1.2.2. Single Crystal X-ray Diffraction

Compared to neutron and electron techniques, single crystal X-ray diffraction is more widely used. It has become a standard analytical method in chemistry to obtain precise insight into structures of compounds. Hereby, X-rays are only diffracted at the electron shells of the atoms and the observed intensities are proportional to the electron number of the corresponding elements. Problems only occur if the number of electrons is very small (e. g. hydrogen atoms) or neighbouring atoms of the periodic table have to be distinguished. In the latter case the connectivity determined by the X-ray experiment can be decisive. In both cases high resolution X-ray or neutron diffraction experiments can be applied to gain the required information.

The remaining question is, why crystals are required for structure determination purposes. Crystals are highly ordered and compose of a large number of equal unit cells (usually 10^{13} to 10^{18}) in a three dimensional translation periodic lattice arrangement. Diffraction is only observed if the wavelength (λ) of the radiation is within the magnitude of the interatomic distances. Thus X-rays (usually $0.5 < \lambda < 2.3$ Å) have to be used. A reflection occurs if the X-ray beam gets diffracted at imaginary planes through equal atomic positions of all unit cells, following Bragg's equation.

$$n\lambda = 2d \sin \theta$$

An nth order reflection can be detected at an angle of θ from lattice planes with distances d between them using radiation of the wavelength λ. These planes are described by Miller indices (hkl) in relative orientation to the axes of the unit cell. The position and intensity of the reflections is used to obtain the electron density within the unit cell by mathematic methods from which the atomic positions can be derived.

2. Research Objectives

Synthetic Aspects

As mentioned in the introduction, the hydrogen-substituted 13/15-parent compounds could only be isolated in the form of their LA/LB-stabilized derivatives. Starting from the phosphanylalanes the higher homologues are still reactive and can undergo hydrogen elimination under formation of different oligomers. Based on the results of *Vogel*, *Schwan* and the results of my diploma thesis it should be possible to control the oligomerization reactions by changing the reaction conditions. Hereby, the polarity of the solvents, temperature and the influence of different Lewis bases are decisive parameters for controlling this process. The final target was to gain insight into the mechanism of the oligomerization using experimental as well as theoretical methods.

Former investigations in our research group report the first Lewis acid free exclusively hydrogen substituted phosphanyl- and arsanylboranes. Within this work it should be figured out if the higher derivatives containing antimony and bismuth could also be obtained.

Crystallography

As the second part of the thesis, the analytical method of single crystal X-ray experiments should be learned and employed to determine structures for scientific co-workers of the Inorganic Department of the University of Regensburg. This included technical services at the diffractometers, the crystal-dependent sample handling, data processing, structure solution and refinement. The main focus was to determine challenging structures starting from space group problems over twinned and disordered ones, and the combination of both, to modulated ones.

3. Synthetic Section

3.1. Isomerization of LA/LB-stabilized Phosphanylalanes

During my diploma thesis [{(CO)$_5$W}H$_2$PAlH$_2$·NEt$_3$] (**1**) was synthesized, which exclusively undergoes a dimerization towards [{(CO)$_5$W}HPAlH·NEt$_3$]$_2$ (**2**) (Equation (15)).[46]

$$2 \quad \underset{\underset{\text{NEt}_3}{\mathbf{1}}}{\overset{(OC)_5W}{H_2P\text{---}AlH_2}} \quad \xrightarrow[-2\,H_2]{CH_2Cl_2,\ r.t.,\ 1\,h} \quad \underset{\mathbf{2}}{\overset{\underset{HP\text{---}AlH}{(OC)_5W\ \ \ \ NEt_3}}{\underset{Et_3N\ \ \ \ W(CO)_5}{HAl\text{---}PH}}} \quad (15)$$

Compound **2** shows interesting behaviour in solution. The ^{31}P{^{1}H} NMR spectrum of a solution of **2** in CD$_2$Cl$_2$ shows three signals. Additionally, in the proton NMR spectrum, resonances for different amine ethyl groups are detected, which indicate the presence of different isomers in solution.

The isomer determined by X-ray structure analysis shows similar substituents to be mutually *cis*. All efforts to isolate different isomers in the solid state failed. All cell parameters for ca. 100 tested crystals are equal within the standard deviation range. Among them, the seven most deviating crystals were processed in full experiments and resulted in the structure of **2** (Figure 2). If these tested crystals are dissolved to record ^{31}P NMR spectra the same signals occur in constant integral ratios. Those suggest a fast equilibrium of isomers present in solution. As *Vogel* reported, an excess of NMe$_3$ leads to the formation of the double-amine-substituted aluminium [{(CO)$_5$W}H$_2$PAlH$_2$·(NMe$_3$)$_2$].[51] Since free amine is present in the reaction mixture, probably due to decomposition, this might cause the isomerization via an intermediate carrying two amine bases. This possibility will be discussed later on.

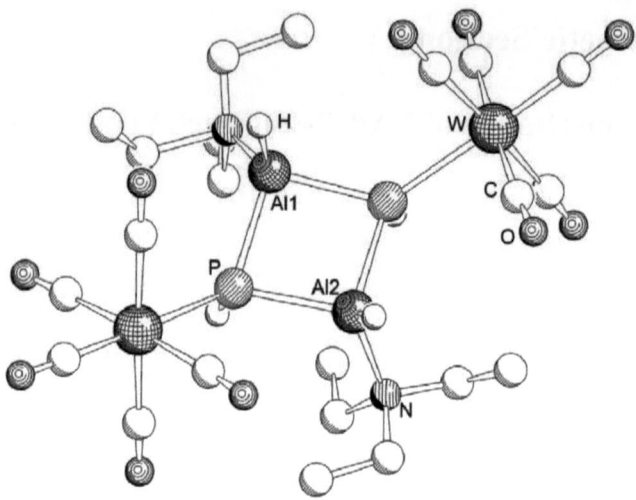

Figure 2: Puckered four-membered ring of **2** in solid state.[46] Hydrogen atoms of the triethylamine are omitted for clarity. Selected bond lengths [Å] and angles [°]: P–Al1 2.382(4), P–Al2 2.385(4), Al1–P–Al2 84.15(15), P–Al1–P 94.62(18), P–Al2–P 94.45(18), Al1–P–Al2–P 12.23(19), mean planes angle (Al1–P–Al2)–(Al2–P–Al1) 16.77.

Influence of a Phenyl Substituent at the Phosphane

The reaction of the phenyl substituted phosphane [{(CO)$_5$W}PPhH$_2$] with H$_3$Al·NMe$_3$ also results in the formation of a four-membered Al$_2$P$_2$ ring compound **3** as the only isolatable product (Equation (16)).[46] The ^{31}P NMR spectrum of the crude reaction mixture shows signals for different isomers present in solution at –138, –140, –142 and –144 ppm. An additional doublet for the monomeric intermediate is observed in the ^{31}P NMR spectrum. Efforts to isolate the monomeric intermediate in solid state failed so far.

(16)

3. SYNTHETIC SECTION

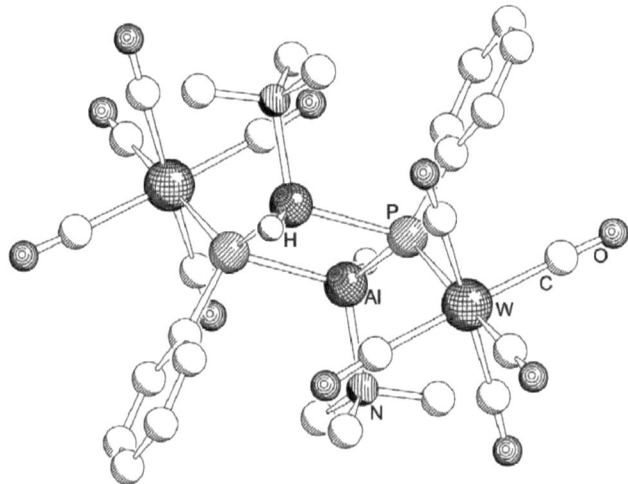

Figure 3: Planar four-membered ring of **3** in solid state.[46] Carbon bound hydrogen atoms are omitted for clarity. Selected bond lengths [Å] and angles [°]: P–Al 2.418(3), Al–P–Al 82.81(9), P–Al–P 97.20(9).

An X-ray structure analysis of **3** reveals a planar four-membered ring motif (Figure 3) with a crystallographic inversion centre in the middle of the Al_2P_2 ring. In contrast to **2**, the determined isomer of **3** shows similar substituents in a *trans* arrangement.

Theoretical Studies[52]

DFT computations were carried out to clarify the mechanisms of formation of **2** and **3**, and to determine the energy differences between their possible isomers (Table 1). The process of the first H_2 elimination from the starting materials (Reactions *1*) is energetically favourable. In contrast, the elimination of the second hydrogen molecule leading to the formation of the hypothetical monomeric [{(CO)$_5$W}HPAlH·NEt$_3$] and [{(CO)$_5$W}(Ph)PAlH·NMe$_3$] (Reactions *2*) is unfavourable, but compensated by the subsequent dimerization energies (Processes *3*). The overall reaction (Processes *5*) is exothermic and thermodynamically allowed both for **2** and **3**. The comparison of the reactions *4a* and *4b* shows, that compound **1** is thermodynamically unstable towards H_2 evolution, while an analogous hydrogen elimination from the [(CO)$_5$WP(Ph)HAlH$_2$·NMe$_3$] monomer is endergonic. The experimental isolation of the more reactive **1** in the solid state during my diploma thesis may be attributed to an additional hydrogen bridge stabilization (Al–H···Al), as reported for

[{(CO)$_5$W}PH$_2$AlH$_2$·NMe$_3$]$_2$.[35,46] Analogous self-dimerization of the phenyl substituted derivative appears to be less favourable due to steric hindrance.

Table 1: Thermodynamic parameters for the reactions leading to the found isomers of **2** and **3**. Predicted standard enthalpies and standard Gibbs energies [kJ mol^{-1}] for gas phase reactions.

	Process	$\Delta H°_{298}$	$\Delta G°_{298}$
1a	H$_3$Al·NEt$_3$ + [{(CO)$_5$W}PH$_3$] = [{(CO)$_5$W}PH$_2$AlH$_2$·NEt$_3$] + H$_2$	−39.1	−24.9
1b	H$_3$Al·NMe$_3$ + [{(CO)$_5$W}P(Ph)H$_2$] = [{(CO)$_5$W}PPhAlH$_2$·NMe$_3$] + H$_2$	−33.4	−24.3
2a	[(CO)$_5$WPH$_2$AlH$_2$·NEt$_3$] = [{(CO)$_5$W}PHAlH·NEt$_3$] + H$_2$	77.0	49.6
2b	[{(CO)$_5$W}PPhHAlH$_2$·NMe$_3$] = [{(CO)$_5$W}PPhAlH·NMe$_3$] + H$_2$	99.9	78.9
3a	[(CO)$_5$WPHAlH·NEt$_3$] = ½ [{(CO)$_5$W}PHAlH·NEt$_3$]$_2$	−97.9	−75.2
3b	[(CO)$_5$WPPhAlH·NMe$_3$] = ½ [{(CO)$_5$W}PPhAlH·NMe$_3$]$_2$	−95.0	−68.8
4a	[{(CO)$_5$W}PH$_2$AlH$_2$·NEt$_3$] = ½ [{(CO)$_5$W}PHAlH·NEt$_3$]$_2$ + H$_2$	−20.9	−25.5
4b	[{(CO)$_5$W}PhHPAlH$_2$·NMe$_3$] = ½ [{(CO)$_5$W}PPhAlH·NMe$_3$]$_2$ + H$_2$	4.9	10.1
5a	H$_3$Al·NEt$_3$ + [(CO)$_5$WPH$_3$] = ½ [{(CO)$_5$W}PHAlH·NEt$_3$]$_2$ + 2 H$_2$	−60.0	−50.4
5b	H$_3$Al·NMe$_3$ + [{(CO)$_5$W}PPhH$_2$] = ½ [{(CO)$_5$W}PPhAlH·NMe$_3$]$_2$ + 2 H$_2$	−28.5	−14.2

Theoretical computations revealed that all five of the possible isomers of **2** (Figure 4) are close in energy, with isomer **ii** being the lowest in energy (Table 2). However, the isomer **i** is the only one that could be characterized by X-ray structure determination (Figure 2).

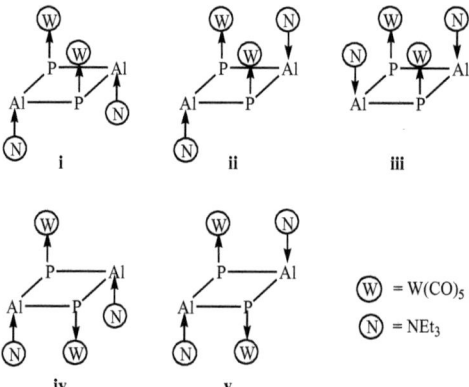

Figure 4: Possible isomers of **2** and **3**. Hydrogen atoms and phenyl groups are omitted for clarity.

3. SYNTHETIC SECTION

Table 2: Relative energies $E°_0$ [kJ mol^{-1}], standard isomerization Gibbs energies $\Delta G°_{298}$ [kJ mol^{-1}] starting from **ii**, dipole moments μ, Debye and dihedral P–Al–P–Al angles $\theta_{(PAlPAl)}$ [°] for isomers of **2**. Values in parentheses correspond to [{(CO)$_5$W}HPAlH·NMe$_3$]$_2$ isomers.

Isomer	$E°_0$	$\Delta G°_{298}$	μ	$\theta_{(PAlPAl)}$
i	3.9 (3.9)	–3.3 (–4.0)	5.7 (6.1)	29.9 (28.0)
ii	0.0 (0.0)	0.0 (0.0)	2.5 (1.9)	17.3 (17.8)
iii	11.3 (12.1)	20.0 (15.7)	4.0 (3.4)	2.8 (1.5)
iv	7.4 (7.5)	9.5 (3.7)	3.2 (1.9)	21.7 (24.1)
v	3.4 (3.4)	5.9 (0.0)	6.2 (4.2)	15.8 (11.4)

The three isomers **i**, **ii**, and **v** are very close in energy (Table 2), which fits well with the experimentally observed equilibrium between isomers in solution. The negative value of isomerization Gibbs energy from isomer **ii** to isomer **i** indicates the latter to be the dominant one in the gas phase at 298 K (assuming the equilibrium between all isomers is achieved). The fact that only isomer **i** is isolated in solid state, probably originates from the favourable packing in the crystal. It possesses a relatively large dipole moment, which increases the crystal lattice energy and facilitates its crystallization. The energy difference between similar NEt$_3$ and NMe$_3$ isomers is very small. In contrast, the substitution of the hydrogen atom at the phosphorus atoms by a phenyl group leads to larger energy differences (up to 25 kJ mol^{-1}) between the five possible isomers of **3** (Figure 4), with the structurally characterized isomer **v** being by 9 kJ mol^{-1} more stable than **iv** (Table 3)

Table 3: Relative energies $E°_0$ [kJ mol^{-1}], standard isomerization Gibbs energies $\Delta G°_{298}$ [kJ mol^{-1}], dipole moments μ, Debye and dihedral P–Al–P–Al angles $\theta_{(PAlPAl)}$ [°] for isomers of **3**.

Isomer	$E°_0$	$\Delta G°_{298}$	μ	$\theta_{(PAlPAl)}$
i	25.7	27.7	13.5	18.6
ii	10.8	16.9	14.2	17.8
iii	15.6	15.8	6.5	24.9
iv	9.1	7.9	4.4	18.3
v	0.0	0.0	0.0	0.0

Gibbs energy values for the isomerization reactions suggest that isomer **v** is the dominant form at room temperature. In agreement with theoretical predictions, the solid state structure determination revealed that isomer **v** features a planar four-membered Al_2P_2 ring (Figure 3). In contrast, the optimized structures of isomer **v** of compound **2** and its NMe_3 analogue are asymmetric with a puckered Al_2P_2 ring. This suggests that the phenyl groups at the phosphorus atoms induce the planarity of the Al_2P_2 ring. A unique feature of the isomer **v** of **3** is the presence of short intramolecular Al–H···H–C contacts of 2.217 Å between negatively charged hydridic hydrogen atoms connected to the aluminium atom and the positively charged hydrogen atom of the phenyl group (Mulliken partial charges are -0.17 and +0.13, respectively). Such interactions can be responsible for the stabilization of isomer **v** in solid state.

Planar rings of element 13/15-compounds are quite common for structures of dimeric imino compounds.[41] For the dimers of heavier group 15 elements both puckered and planar structures are observed, with planar structures usually enforced by very bulky substituents. Thus, the structurally characterized donor-only-stabilized aluminium-phosphorus dimer $[(^iPr_3Si)PAlClPy]_2$ has a planar Al_2P_2 ring.[53]

The remaining question is what causes the isomerization. The experimentally observed equilibrium between isomers of **2** in solution together with the presence of free amine indicates a fast ligand exchange process. Computational data (Table 1) reveal that dimers are quite strongly bound with respect to dissociation of Al_2P_2 ring into monomers. For **2(ii)** and **3(v)** such a dissociation is endothermic by 190-196 kJ mol^{-1} (per mole of dimer). These results suggest that this process cannot be responsible for the fast isomerization of **2** at room temperature. As reported earlier, an amine exchange is possible for an intermediate of a trigonal bipyramidal aluminium atom carrying two amine ligands in presence of free amine.[51] In the case of **2**, such an exchange of one NEt_3 ligand would lead to the isomers **ii** and **iii** resulting in the observed additional resonances. However, two base-exchange mechanisms are possible (Scheme 4).

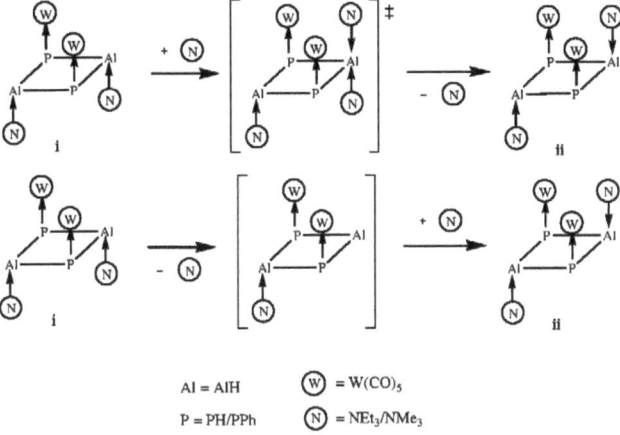

Scheme 4: Proposed S_N2 (top) and S_N1 (bottom) isomerization pathways from **i** to **ii**.

Depending on the isomer, dissociation of NEt$_3$ from **2** is endothermic by 107–132 kJ mol^{-1}, and dissociation of NMe$_3$ from [{(CO)$_5$W}PhPAlH·NMe$_3$]$_2$ is also unfavourable by 95–135 kJ mol^{-1}. In both cases, the most stable product [{W(CO)$_5$P}RAlH$_2$NR'$_3$] (R = Ph, H; R' = Me, Et) features the W(CO)$_5$ group in a bridging position (Figure 5).

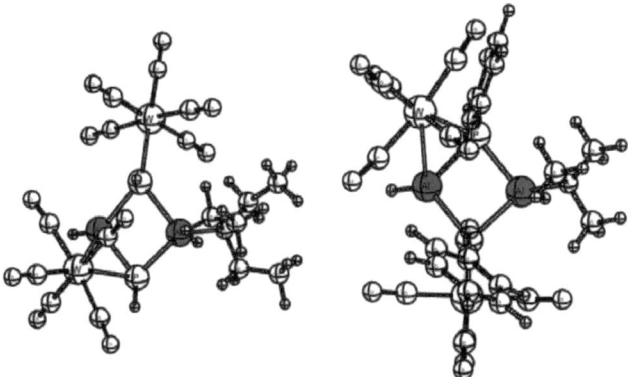

Figure 5: Optimized intermediates of the base abstraction reaction of **2** (left) and **3** (right).

The reaction energies of the base dissociation processes are the lower limits for the activation energies of the S_N1 mechanism. These values (95–135 kJ mol^{-1}) appear to be too high to account for the observed quick isomerization at room temperature. Thus, isomerization via an S_N1 mechanism with amine dissociation can be ruled out as energetically demanding. The alternative S_N2 pathway includes the addition of the amine to the dimeric ring and appears to be more probable. However, attempts to optimize intermediate structures for the S_N2 mechanism (by addition of the amine to **2** and **3**, respectively) failed. Proposed intermediates with five-coordinate trigonal pyramidal aluminium proved to be unstable. Such structures eliminate one amine upon optimization. This indicates an easy removal of the amine from the intermediate, but does not allow determination of the activation energy for the first stage of the S_N2 reaction pathway.

Hence, the most probable conclusion is a fast base exchange via a S_N2 mechanism, both for **2** and **3**. Furthermore, the presence of the free amines in solution supports this theory. Their signals in the ^1H NMR spectra are broad, which provides a further hint for a dynamic process.

3.2. The Controlled Oligomerization of [{(CO)$_5$W}H$_2$PAlH$_2$·NMe$_3$] (4)[54]

As reported by *Vogel*, **4** is formed by a hydrogen elimination reaction of [{(CO)$_5$W}PH$_3$] with H$_3$Al·NMe$_3$ in refluxing *n*-pentane and gives yields of 45%.[35] This reaction can be improved by changing the solvent to dichloromethane, with an increased yield of 82%. Interestingly, a solution of **4** in CH$_2$Cl$_2$ shows further reactivity, liberating additional equivalents of H$_2$, even at room temperature. For this reason a variety of different compounds could be obtained from the starting material **4**, and extensive investigations allowed a clarification of the reaction mechanism. In contrast to the dimerization of its triethylamine derivative **1**, *Vogel* found the trimerization product and NMR evidence for further compounds, but those have not been further characterized.[45] However, a low quality X-ray structure of the trimer (**5**) was described.

Trimerization of 4 to yield [{(CO)$_5$W}HPAlH·NMe$_3$]$_3$ (5)

If crystals of **4** are dissolved in toluene and warmed to 30 °C, further hydrogen evolution is observed. Yellow crystals of [{(CO)$_5$W}HPAlH·NMe$_3$]$_3$ (**5**) are obtained in 47% yield as the only isolatable product (Equation (17)).

(17)

Compound **5** is only poorly soluble in dichloromethane and decomposes in coordinating solvents like THF. If the reaction mixture of [{(CO)$_5$W}PH$_3$] and H$_3$Al·NMe$_3$ in dichloromethane is not cooled to –28 °C after the first hydrogen elimination step, but kept at room temperature, crystals of **5** are obtained in moderate yields (22%) by liberating another equivalent of hydrogen, in addition to crystals of [⟨{(CO)$_5$W}HPAlH·NMe$_3$⟩$_2$⟨(CO)$_5$WPAl·NMe$_3$⟩] (**6**) (Scheme 5).

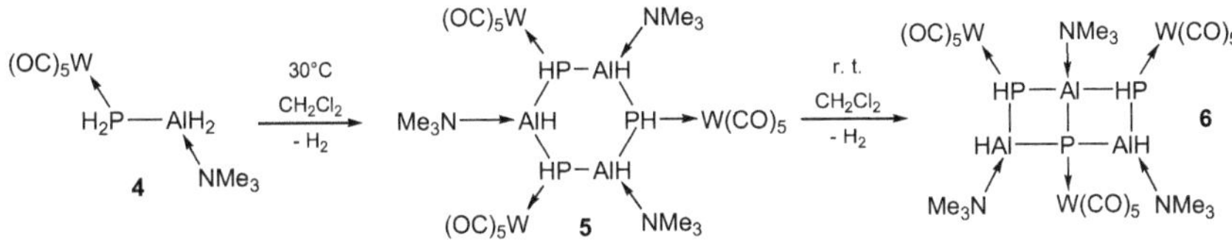

Scheme 5: Reaction pathway of the trimerization of **4**.

Due to the low solubility of **5**, overnight NMR experiments had to be carried out in CD$_2$Cl$_2$. During these experiments, additional signals were found, which were identified as signals of **6**. The ^{1}H NMR spectrum of **5** gives three different broad doublets for the phosphorus-bound protons at 0.26 (^1J(HP) = 242 Hz), 0.51 (^1J(HP) = 238 Hz) and 0.54 (^1J(HP) = 223 Hz), along with two singlets for the NMe$_3$ protons at 2.84 and 2.86 ppm in an integral ratio of 2:1. A phosphorus decoupled proton NMR spectrum merges all three PH doublets to singlets, proving that all phosphorus atoms carry hydrogen substituents. The ^{31}P NMR spectrum shows the expected doublets at –328.5, –328.2 and –317.4 ppm with the corresponding coupling constants of the ^{1}H NMR spectrum. The proton signals as well as those in the ^{31}P NMR spectrum are shifted upfield compared to those of the six-membered ring [(^{i}Pr$_3$Si)P(H)AlMe$_2$]$_3$ (^{1}H NMR: 1.15 ppm, ^{31}P NMR –241 to –252 ppm).[55] This fact can be explained by its different Al–P bonding. In this compound the donor and acceptor functions of the phosphorus and aluminium atoms are unblocked compared to those in **5**. This leads to additional bonding interactions and better shielding of the NMR active nuclei.

The IR spectrum of **5** shows absorptions at 2299 and 2276 cm^{-1} for P–H and at 1670 cm^{-1} for the Al–H vibrations. Additionally, carbonyl bands are observed at 2079, 2062 and 1915 cm^{-1}. Those are in good agreement with the theoretically computed values (Figure 6).[52]

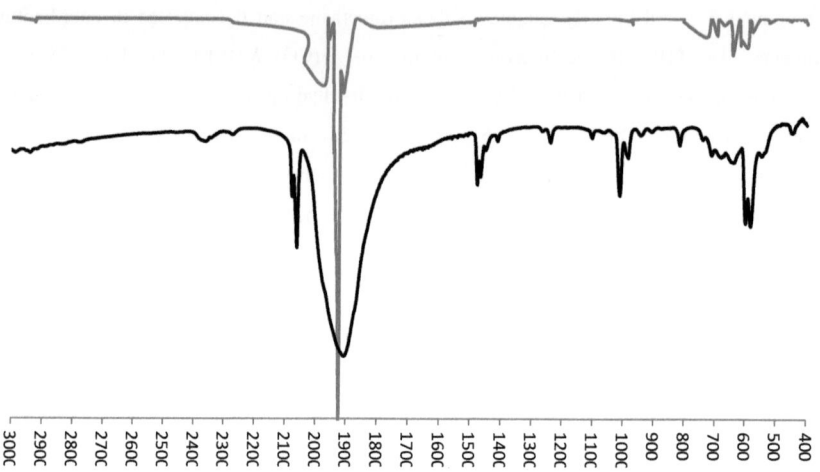

Figure 6: Experimental (lower) and computed (upper) IR spectra of **5** (wave numbers are given in cm^{-1}). Scaled harmonic vibrational frequencies (ν) according to Equation $\nu = 0.9461\omega + 22.1$.[56]

A fragment of the Al$_3$P$_3$H$_6$(NMe$_3$)$_3$ core can be found in the mass spectrum at m/z = 357, but is overlapped with characteristic fragments of [{(CO)$_5$W}PH$_3$] (m/z = 358) and its successive CO elimination ions. These cannot be avoided due to the sample preparation procedure. Thereby a short exposure to air is necessary and leads to a reaction with traces of moisture resulting in a cleavage of the P–Al bond forming the observed Lewis-acid-stabilized phosphane and an OH$^-$ addition at the aluminium.

An X-ray structural analysis of **5** shows a distorted six-membered Al$_3$P$_3$ ring in the boat conformation (Figure 7). Each of the ring atoms carries one hydrogen substituent. The atoms P1 and P2 are coordinated by [W(CO)$_5$] units in equatorial and at P3 in axial position. NMe$_3$ also coordinates in two different ways. The amine bases at the atoms Al1 and Al2 adopt equatorial positions and an axial position at Al3.

Interestingly, for the parent compound, Al$_3$P$_3$H$_6$, the C_{3v} symmetric chair conformation is predicted to be energetically favoured by 16 kJ mol^{-1} with respect to the C_s symmetric boat conformer.[57] In contrast, the calculated energies for the isomers of **5** differ by less than 7 kJ mol^{-1}, with a structure corresponding to the experimentally observed conformer (**II**) being the most stable (Figure 8).[52]

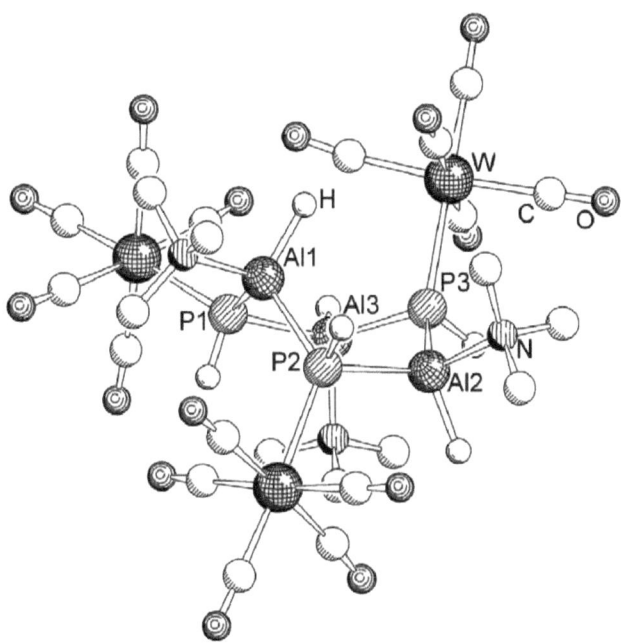

Figure 7: Molecular structure of **5** (methyl hydrogen atoms are omitted for clarity). Selected bond lengths [Å] and angles [°]: P1–Al1 2.366(2), P1–Al3 2.383(2), P2–Al1 2.375(2), P2–Al2 2.369(2), P3–Al3 2.368(2), P3–Al3 2.362(2), Al1–P1–Al3 112.63(5), Al1–P2–Al2 103.88(6), Al2–P3–Al3 110.60(6), P1–Al1–P 111.63(6), P2–Al2–P3 112.00(6), P1–Al3–P3 112.53(6).

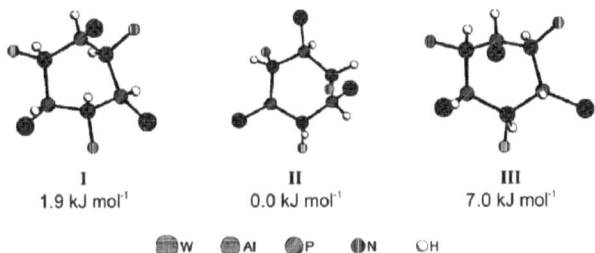

Figure 8: Comparison of the structures and relative energies for the considered possible isomers of **5**. Carbonyl and methyl groups omitted for clarity.

Compound **5** is the first example of a LA/LB-stabilized phosphanylalane oligomer forming exclusive σ-bonds between the group 13 and 15 element (in comparison to partial dative π-bonding interactions in donor/acceptor unblocked systems). This fact is clearly revealed in the

Al–P bond lengths (2.362(2)–2.383(2) Å). Those are shorter than e.g. in the previously mentioned trimer [(^{i}Pr$_3$Si)P(H)AlMe$_2$]$_3$ (2.453(2) Å), in which additional donor-acceptor bonds exist.[55] Consequently the Al–P bond lengths in **5** are in good agreement with an exclusive σ-bond as in comparable monomers such as (Me$_3$Si)$_2$PAlMe$_2$·dmap (2.379(1) Å).[58] However, the Al–P bonds are longer compared to those in (Mes*AlPPh)$_3$ (2.323(3)–2.336(3) Å), in which additional weak π-interaction is expected.[59] All bond angles in **5** within the ring are almost equal (110.60(6)–112.63(6)°) except for the Al1–P2–Al2 angle, which is compressed to 103.88(6)°. This fact is a consequence of the arrangement of the large W(CO)$_5$ units, and leads to the distortion of the ring.

Further Reactivity of 5 under Formation of the Ladder Compound [〈{(CO)$_5$W}HPAlH·NMe$_3$〉$_2$〈(CO)$_5$WPAl·NMe$_3$〉] (6)

As mentioned before a further product can be obtained from the reaction of **4** in dichloromethane at room temperature. Besides the six-membered trimer **5** crystals of the ladder-like compound [〈{(CO)$_5$W}HPAlH·NMe$_3$〉$_2$〈(CO)$_5$WPAl·NMe$_3$〉] (**6**) could be isolated. Compound **6** could also be generated if crystals of **5** obtained from the reaction in toluene are re-dissolved in dichloromethane and treated with ultrasound at ambient temperature (Scheme 6). During this procedure an additional equivalent of hydrogen is eliminated. However, the reaction of **5** to **6** is incomplete before decomposition occurs. Hence, crystals of **6** were isolated by separating them from those of **5** by means of different colour and shape.

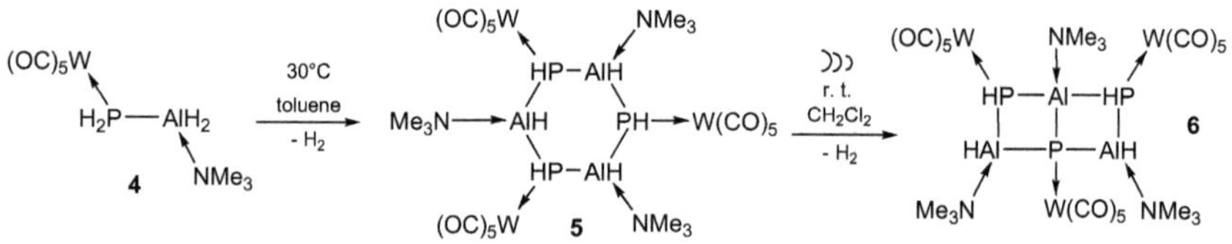

Scheme 6: Controlled trimerzation of **4** corresponding to the reaction conditions.

Evidence for **6** could already be seen in the NMR experiments of **5**, since the polar solvent CD$_2$Cl$_2$ had to be used. Compound **6** is even less soluble in dichloromethane than **5**, and a precipitate of **6** was observed in the NMR tube. The ^{1}H NMR spectrum shows two doublets for the phosphorus-bound protons at 0.18 (d, ^1J(HP) = 234 Hz) and 0.90 ppm (d, ^1J(HP) = 227 Hz) which become singlets in the phosphorus decoupled spectrum. In contrast to **5**, only one broad singlet for the three amine bases was observed. In the ^{31}P NMR spectrum two doublets at –289.4 and –267.6 ppm are detected for the hydrogen substituted phosphorus atoms. The singlet for the phosphorus atom connected to three aluminium atoms is observed at −312.3 ppm. The low field shift compared to the doublets of **5** can be explained by the compression of the core atom angles and its corresponding changes in the electron distribution in the relevant orbitals. For this reason [{(CO)$_5$W}HPAlH·NEt$_3$]$_2$ with a four-membered ring core shows similar chemical shifts at −276.4, −282.6 and −283.2 ppm (different isomers) and coupling constants (232, 233 and 234 Hz).[46] In comparison, the singlet is high field shifted due to four bonds towards three aluminium and one tungsten atoms, all of which are strong Lewis-acids, leading to an additional deshielding.

The mass spectrum shows an ion peak for [Al$_3$P$_3$H$_4$(NMe$_3$)$_3$]$^+$ at 355 *m/z* only slightly shifted to [{(CO)$_5$W}PH$_3$] at 358 *m/z*.

The IR spectrum of **6** in solid state shows a P–H stretch at 2274 cm^{-1}, an Al–H stretch at 1647 cm^{-1} and carbonyl stretches in the characteristic range (2078 and 1927 cm^{-1}). As in the case of **5** a theoretical spectrum was computed and shows a good agreement with the experimental one (Figure 9).[52]

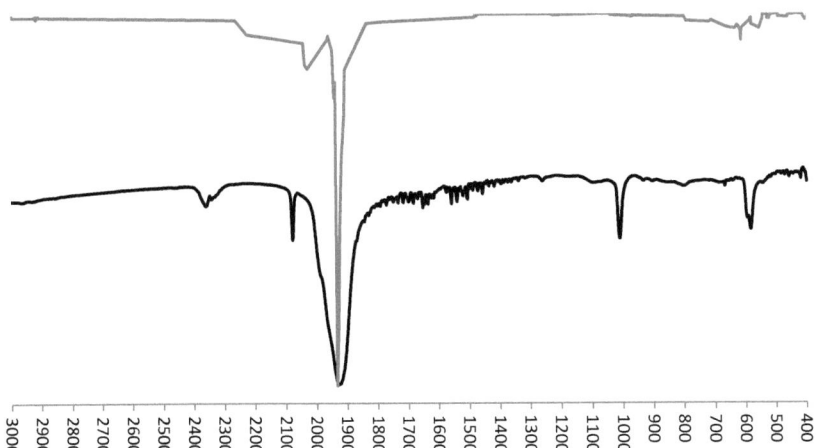

Figure 9: Experimental (upper) and computed (lower) IR spectra of **6** (wave numbers are given in cm^{-1}). Scaled harmonic vibrational frequencies (*v*) according to equation $v = 0.9461\omega + 22.1$.[56]

The X-ray structural analysis of **6** shows a distorted Al_3P_3 ladder core (Figure 10). The aluminium and phosphorus atoms at the corners (Al1, P1, Al2, P3) still carry hydrogen substituents, whereas, the central atoms Al3 and P2 bind to each other, which is a result of the H_2 elimination in comparison to the *cyclo*-trimer **5**.

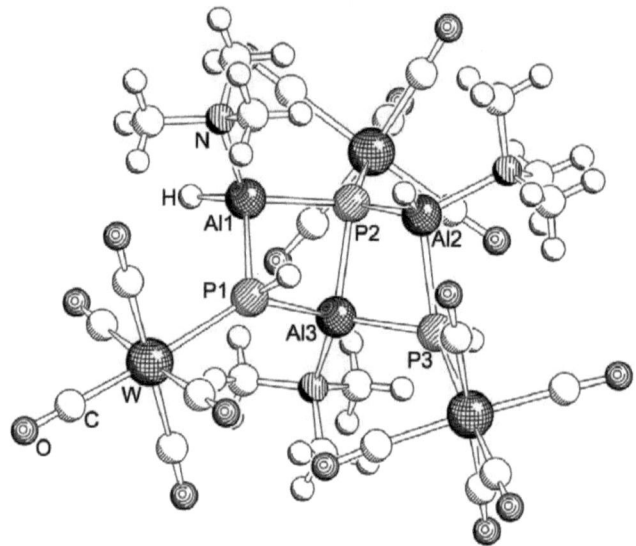

Figure 10: Molecular structure of **6** in the crystal. Selected bond lengths [Å] and angles [°]: P1–Al1 2.382(2), P1–Al3 2.365(3), P2–Al1 2.333(3), P2–Al2 2.386(3), P2–Al3 2.394(3), P3–Al2 2.385(2), P3–Al3 2.332(3), Al1–P1–Al3 79.16(8), Al1–P2–Al2 106.2(1), Al1–P2–Al3 79.55(9), Al2–P2–Al3 76.83(8), Al2–P3–Al3 78.04(8), P1–Al1–P2 97.67(9), P2–Al2–P3 96.29(9), P1–Al3–P2 96.48(9), P2–Al3–P3 97.51(9), P1–Al3–P3 113.5(1).

The P–Al bond lengths of **6** (2.332(3)–2.394(3) Å) correspond to single bonds and are in good agreement to those in **5**. In contrast to the six-membered ring structure in **5**, most of the angles in **6** are strongly bent to values below 100° (97.67(9)–76.83(8)°) as a result of the strained bicyclic ring substructures (Al1–P1–Al3–P2 and Al2–P2–Al3–P3). Other four-membered Al_2P_2 rings like $(Ph_2PAl^iBu_2)_3$ also show small angles (Al–P–Al 93.8(1) and P–Al–P 86.2(1)°), but the differences within the corresponding ring systems are smaller due to the longer Al–P bond lengths (2.475(1) Å) caused by the mixture of donor-acceptor and non-dative bonding interactions.[60] The range of bond lengths and angles is also in good agreement with those reported for an eight-membered ladder compound $[ClAlPR]_4 \cdot 2\ Et_2O$ (R = Si^iPr_3 and $SiMe^iPr_2$) (2.280(1)–2.427(1) Å and 78.51(5)–121.76°). However, in this compound again donor-acceptor bonding interactions are proposed. Furthermore, the electron withdrawing influence of the chlorine substituents effects the bond lengths.[44]

Theoretical and mechanistic Considerations[52]

Among ten considered isomers of the ladder compound **6**, the structure corresponding to the one found experimentally is the most stable (**X**) (Figure 11).[54] The maximum energy difference between the isomers is 29 kJ mol^{-1}. For the parent (LA/LB-free) Al$_3$P$_3$H$_4$ ladder, the maximal difference between conformers is 25 kJ mol^{-1}. The C_s symmetric structure with a *cis*-orientation of the lone pairs at the phosphorus atoms is the most stable. The structures of the most stable parent (LA/LB-free) and LA/LB-stabilized ladder **6** do not match. Thus, the influence of the sterics of the LA and LB dictates the formed ladder conformer as well.

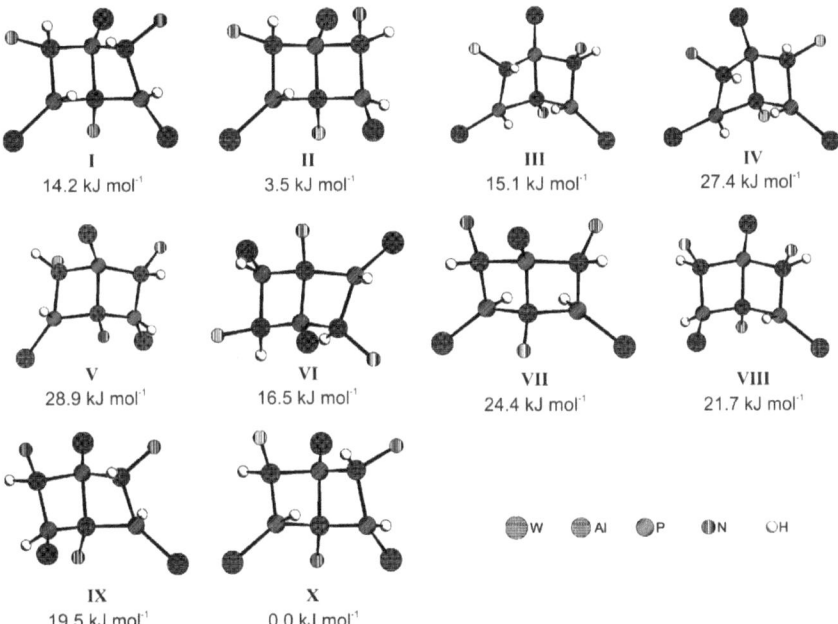

Figure 11: Comparison of the structures and relative energies for the considered possible isomers of **6**. Carbonyl and methyl groups omitted for clarity.

According to the oligomerization and polymerization of phosphanylboranes investigated by *Manners et al.*, initial approaches using [(COD)Rh(μ-Cl)]$_2$ as hydrogen elimination catalyst were carried out.[24] The presence of the Rh(I) catalyst causes no difference in the reactivity of

4 towards **5** and **6** at ambient temperatures. Nevertheless, besides the two previously described compounds a very minor side-product was identified as [{(CO)$_5$WPH$_2$}(Me$_3$N)AlPH{W(CO)$_5$}]$_2$ (**7**) containing a four-membered Al$_2$P$_2$-ring carrying two additional LA-stabilized phosphane units (Scheme 7).[54]

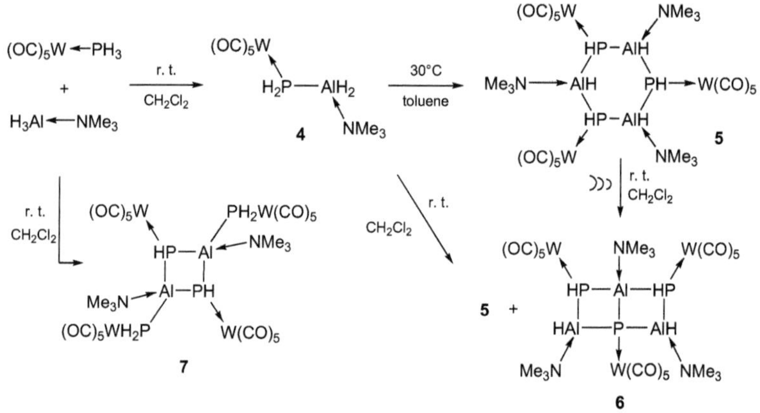

Scheme 7: Overview of all isolatable compounds obtained from the reaction of [{(CO)$_5$W}PH$_3$] with H$_3$Al·NMe$_3$.

In addition to the reported analytical data, NMR chemical shifts could be obtained from the crude reaction mixture for **7**, since the other signals can now be assigned to **5** and **6**. The ^{1}H NMR spectrum of **7** shows a doublet at 0.66 (^1J(HP) = 229 Hz) for the endocyclic PH groups and a triplet at 2.83 ppm (^1J(HP) = 287 Hz) for the exocyclic PH$_2$. Furthermore, the singlet for the amine protons is found at 2.82 ppm. The ^{31}P NMR spectrum of **7** reveals a triplet at –234.6 ppm for the phosphorus atoms outside and a doublet at –287.6 ppm for those within the ring. The chemical shift and coupling constant of the exocyclic PH$_2$ is similar to that of **4** (^{1}H NMR: 1.93 ppm (^1J(HP) = 283 Hz), ^{31}P NMR: –250 (^1J(PH) = 283 Hz)),[35] whereas those of the endocyclic PH are closer to those of **6**.

As mentioned before, the ladder compound **6** is formed together with the *cyclo*-trimer **5** from the monomer **4** in CH$_2$Cl$_2$. Experiments prove that direct H$_2$ elimination from **5** leads to **6**. Theoretical studies using DFT calculations for corresponding gas phase reactions support this pathway due to a Gibbs energy of –35.5 kJ mol^{-1} for this reaction.[52] A second pathway for the formation of **5** is further proposed in which two units of **4** form a four-membered ring [{(CO)$_5$W}HPAlH·NMe$_3$]$_2$ (**8**) that adds a third molecule of **4** to result the ladder molecule **6** (Scheme 8).

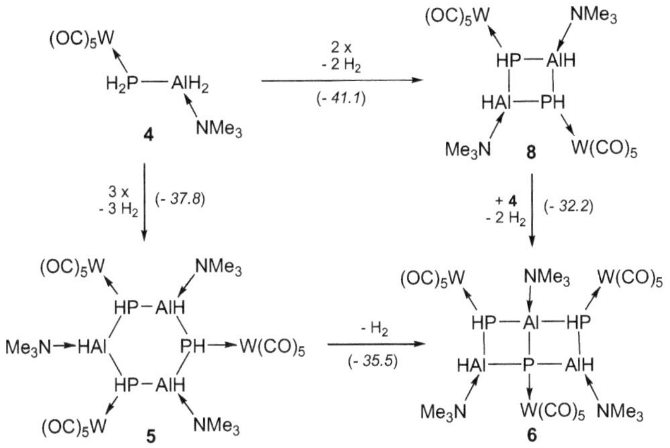

Scheme 8: Different reaction pathways to the formation of **5** and **6** (*Gibbs energies kJ mol^{-1}*).

The computations show the formation of the dimer **8** (–41.1 kJ mol^{-1}) thermodynamically in competition with the generation of the trimerization product **5** (–37.8 kJ mol^{-1}). Intramolecular H$_2$ elimination from **5** resulting in **6** is only slightly more exergonic (–35.5 kJ mol^{-1}) than the addition of the monomer **4** to the dimer **8** (–32.2 kJ mol^{-1}). In fact, stopping the reaction after a short period of time led to ^{31}P NMR evidence for a possible intermediate **8** in the form of a set of doublets at –333.3, –337.3, –340.7 and –349.9 ppm with ^{1}J(PH) couplings of approximately 210 to 270 Hz in the reaction mixture (most likely different isomers as previously described for **2** and **3**). However, attempts to isolate **8** failed so far. Another question is the formation mechanism of the four-membered ring product **7**. Two pathways are possible (Scheme 9).

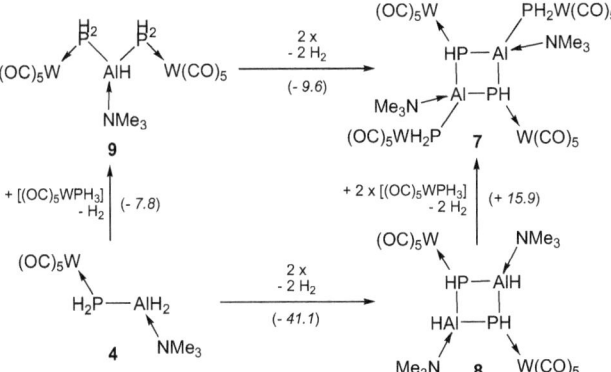

Scheme 9: Different reaction pathways to the formation of **7** (*Gibbs energies kJ mol^{-1}*).

Two equivalents of [{(CO)$_5$W}PH$_3$] could add to **8**, resulting in **7**. Alternatively another molecule of [{(CO)$_5$W}PH$_3$] could add to the monomer **4**, and the so-formed compound [{(CO)$_5$W}PH$_2$]$_2$AlH·NMe$_3$ (**9**) gives **7** via subsequent H$_2$ elimination. As the computations show, the first route (**4** → **8** → **7**) includes the thermodynamically unfavourable step **8** → **7** (+15.9 kJ mol^{-1}), whereas the alternative (**4** → **9** → **7**) pathway includes only favourable steps: The formation of an intermediate **9** from **4** (−7.8 kJ mol^{-1}) with a subsequent dimerization of **9** under H$_2$ elimination to **7** (−9.6 kJ mol^{-1}). In agreement with these assumptions and considering the ^{31}P NMR spectrum of the crude reaction mixture a further triplet at 244 ppm could be assigned to a possible compound **9**. The decomposition of **4** with the formation of solid aluminium and **7** is also a thermodynamically favoured process, and this pathway may contribute to the formation of **7** (Table 4). A similar decomposition has been reported for (Mes)$_2$PAlH$_2$·NMe$_3$ in the solid state upon heating the solid compound to 125 °C, resulting in (Mes)$_2$PH, metallic aluminium, NMe$_3$ and H$_2$.[61] Hereby the first step is a trimerization under NMe$_3$ elimination at 95 °C.

Table 4: Selected thermodynamic data. Reaction energies $\Delta E°_0$, standard enthalpies $\Delta H°_{298}$ and standard Gibbs energies $\Delta G°_{298}$ in kJ mol^{-1}, standard entropies $\Delta S°_{298}$ in J mol^{-1} K^{-1}.

Process	$\Delta E°_0$	$\Delta H°_{298}$	$\Delta S°_{298}$	$\Delta G°_{298}$
1 + **2** = **3** + H$_2$	−34.2	−38.2	−35.7	−27.5
1 + **2** = ½ **7** + 2 H$_2$	−48.8	−58.8	−35.8	−48.1
1 + **2** = ⅓ **4** + 2 H$_2$	−57.4	−65.5	−85.0	−40.1
1 + **2** = ⅓ **5** + 2⅓ H$_2$	−52.1	−63.0	−37.1	−52.0
2 × **1** + **2** = **8** + 2 H$_2$	−59.8	−67.9	−109.3	−35.3
2 × **1** + **2** = ½ **6** + 3 H$_2$	−62.4	−76.6	−122.2	−40.2
4 × **3** = **6** + 2 Al$_{(solid)}$ + 2 NMe$_3$ + 5 H$_2$		−39.1	214.2	−103.0

Employing these results, the synthesis of **7** was optimized to a yield of 27%, by slowly adding a less concentrated solution of one equivalent H$_3$Al·NMe$_3$ in dichloromethane to a more concentrated solution of two equivalents of the phosphane complex in the same solvent.

3.3. A new Cage Motif of Phosphanylalanes employing the Lewis Base NMe$_2$Et

The reaction of [{(CO)$_5$W}PH$_3$] with H$_3$Al·NMe$_2$Et in toluene leads to a few pale yellow crystals of [{(CO)$_5$WPH$_2$}(Me$_2$EtN)AlPH{W(CO)$_5$}]$_2$ (**10**) and [⟨{W(CO)$_5$}HPAl(Me$_2$EtN)⟩$_2$ μ-⟨{(CO)$_5$WPH}$_2$Al(Me$_2$EtN)⟩] (**11**) after six months (Scheme 10). Changes in stoichiometry and different reaction conditions did not result in different products at all. Both compounds appeared to be insoluble even in CH$_2$Cl$_2$, thus an analytical characterization in solution could not be carried out. Furthermore, the crystals of **10** and **11** cannot be separated optically and the reaction only yielded a few crystals. For that reason, and regarding the very long reaction time, no further analytical data can be reported. However, an X-ray crystal determination was successful.

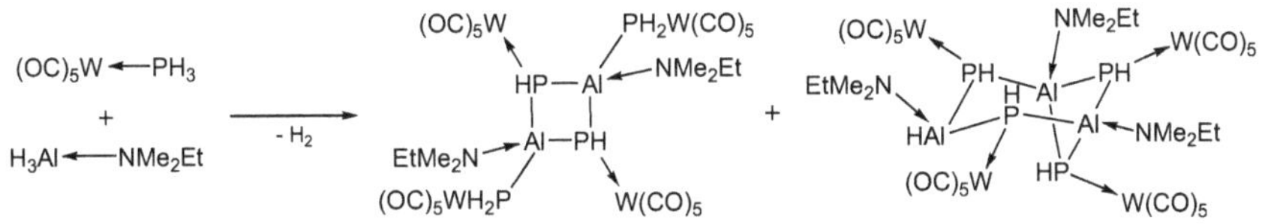

Scheme 10: Synthesis of **10** and **11**.

Compound **10** is the NEt$_2$Me derivative of **7**,[54] containing a four-membered Al$_2$P$_2$ ring motif and two exocyclic tungsten phosphane units, arranged mutually *trans* (Figure 12). As in its NMe$_3$ analogue **7**, the ring is planar (0.00(13)°). The endo- and exocyclic Al–P bond lengths (2.383(3), 2.375(4), 2.371(5) Å) are in the range of those in **5**, **6** and **7**. The angles within the ring (Al–P–Al 80.46(12), P–Al–P 99.54(13)°) are comparable to those in **7** (P–Al–P 98.86(17), Al–P–Al 81.14(17)°) and also to the four-membered ring substructures in **6**. In contrast, the angles containing the exocyclic phosphorus atoms (109.87(14), 108.58(16)°) are in line with those in **5** and the exocyclic ones in **7**. In other previously reported Al$_2$P$_2$ rings such as [(Me$_3$Si)$_2$PAlMe$_2$]$_2$ the angles are closer to 90° (P–Al–P 89.4(3) and Al–P–Al 90.60(5)°) due to longer Al–P bonds (2.460 Å) originating from dative bonding portions.[62]

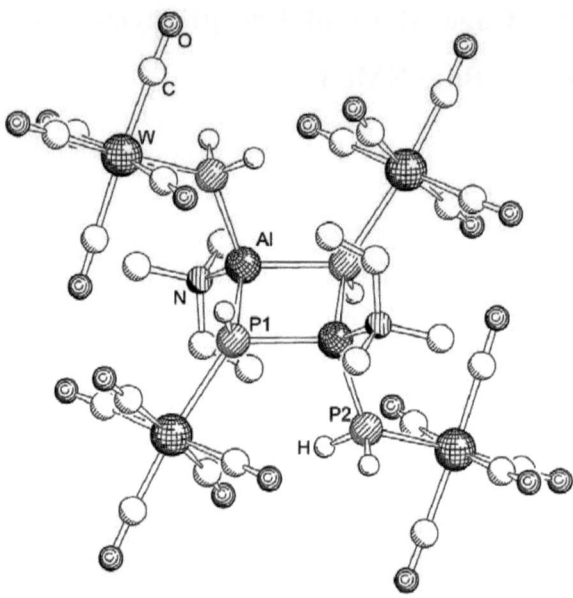

Figure 12: Molecular structure of **10** in the crystal. Selected bond lengths [Å] and angles [°]: P1–Al 2.383(3), P1–Al' 2.375(4), P2–Al 2.371(5), Al–P1–Al' 80.46(12), P1–Al–P1' 99.54(13), P1–Al–P2 109.87(14), P1–Al–P2' 108.58(16).

Compound **11** features a previously unknown structural motif for phosphanylalanes,[63] which can be described in two ways. One is derived from **10**, but with *cis* arrangement of similar substituents. Since both exocyclic phosphanes point to the same side in comparison to **10**, they are now bridged by an additional HAl·NMe$_2$Et unit. The second description is closer to the structure of **5**, because **10** shows bicyclic six-membered ring motifs. The arrangement of the substituents in the four-membered Al1–P1–Al2–P2 substructure leads to a puckered ring with a torsion angle of 30.33(10)°. Another feature the X-ray structure revealed was the disorder in the bridging HAl·NMe$_2$Et unit. The major component with 62% occupancy shows an almost flat arrangement of the Al1–P3–Al3A–P4–Al2 part of the structure (Figure 13). The maximum deviation from the mean plane is only 0.161 Å for P3.

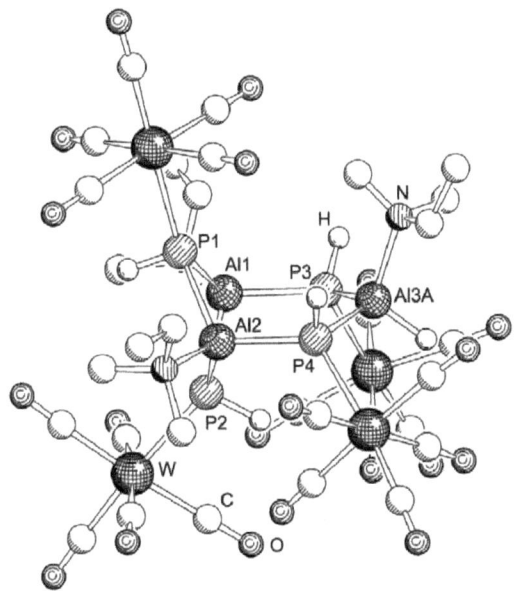

Figure 13: Main part of the disordered crystal structure of **11**. Selected bond lengths [Å] and angles [°]: P1–Al1 2.371(3), P1–Al2 2.361(3), P2–Al1 2.376(3), P2–Al2 2.378(3), P3–Al1 2.349(3), P3–Al3A 2.372(5), P4–Al2 2.362(3), P4–Al3A 2.369(5), Al1–P1–Al2 79.87(11), Al1–P2–Al2 79.42(10), Al1–P3–Al3A 109.84(15), Al2–P4–Al3A 108.66(14), P1–Al1–P2 91.92(11), P1–Al1–P3 120.88(13), P2–Al1–P3 103.88(12), P1–Al2–P4 115.74(12), P2–Al2–P4 112.00(12), P1–Al2–P2 92.12(11), P3–Al3A–P4 115.56(16).

In contrast, in the minor component Al3B is strongly bent with a mean plane deviation of 0.647 Å in this part of the structure (Figure 14). The different disordered aluminium atoms also carry their amine bases in different orientations, Al3A has the amine in an axial, and Al3B in equatorial position. Hence, the structure of **11** allegorises a mixed crystal of two different isomers. Additionally, even the position of the ethyl group of the LB at Al1 is dependent on the Al3 disorder. Furthermore, two carbonyl ligands show different disordered orientations.

The Al–P bond lengths in **11** are in line with those of the previously reported ones (2.349(3)–2.378(3) Å). Only those on the minor aluminium position Al3B are slightly elongated to 2.405(8) and 2.400(7) Å. In comparison to **5**, the minor component reveals strongly bent Al1–P3–Al3B and Al2–P4–Al3B angles (90.42(18), 91.99(18)°). These values are similar to the Al–P–Al angles in the four-membered ring substructure (91.92(11), 92.12(11)°). This is probably caused by steric repulsion of the amine base and the neighbouring carbonyl groups.

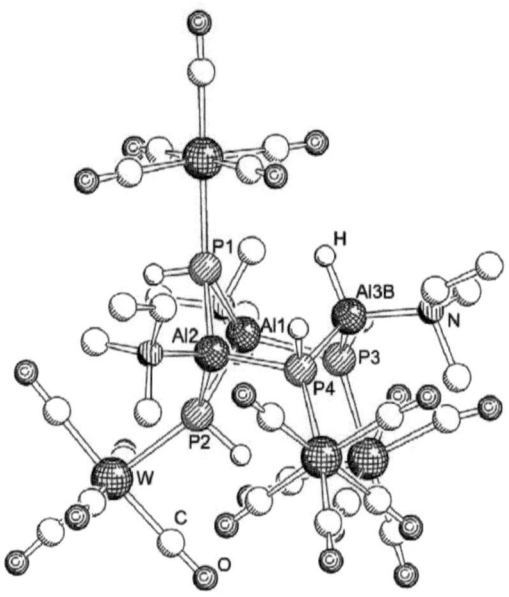

Figure 14: Minor component of the disordered crystal structure of **11**. Additional bond lengths [Å] and angles [°]: P3–Al3B 2.405(8), P4–Al3B 2.400(7), Al1–P3–Al3B 90.42(18), Al2–P4–Al3B 91.99(18), P3–Al3B–P4 113.2(3).

3.4. Lewis acid-free Phosphanylalanes

As mentioned before, no LA-free hydrogen-only-substituted phosphanylalanes analogous to H_2PBH_2·LB have been obtained to date. However, the partially silylated phosphanes show interesting reactivity.

The System $(Me_3Si)_2PH$ and H_3Al·NMe_3

The reaction of $(Me_3Si)_2PH$ with H_3Al·NMe_3 leads to the expected compound $(Me_3Si)_2PAlH_2$·NMe_3 (**12**) (Equation (18)). This hydrogen elimination is observed in dichloromethane and in toluene at room temperature, forming **12** in good yields (ca. 50%).

3. SYNTHETIC SECTION

$$(Me_3Si)_2PH + H_3Al\leftarrow NMe_3 \xrightarrow{-H_2} \underset{\mathbf{12}}{Me_3Si\cdots P-Al(NMe_3)(H)_2 \text{ with } Me_3Si} \tag{18}$$

In addition to the high sensitivity of **12** towards air and moisture, the product is unstable at room temperature and has to be kept in the mother liquor. Decomposition occurs already at ambient temperatures, as has been proven by NMR studies of the crude reaction mixture. From concentrated solutions colourless crystals of **12** are obtained after a few hours at 4 °C, but decomposition is indicated by an additional white precipitate. Efforts to synthesize or crystallize the compound at lower temperatures failed. If the reaction is carried out at temperatures below 0 °C no hydrogen elimination is observed. Crystallization at –28 °C results in no crystals, but the solutions turns milky white in both solvents.

The ^{1}H NMR of the crude reaction mixture in CD_2Cl_2 shows a doublet of multiplets at 0.55 ppm ($^1J(HP)$ = 4.4 Hz) for the methyl groups of the silyl substituents. The singlet at 1.86 ppm gives evidence for the methyl groups of the amine base and a further broad signal for the hydrogen atoms at the aluminium is observed at 4.5 ppm. The ^{31}P NMR resonance occurs at – 283.3 ppm, which is shifted upfield compared to the LA/LB-stabilized compound **4** (–250 ppm).[34] The reason for this is that the absence of the LA $W(CO)_5$ is overcompensated by the two electropositive silyl ligands replacing the hydrogen atoms.

Due to the instability of compound **12** the crystals had to be taken from the cool solution directly to a cooled well containing the perfluorinated polyethers (chapter 4.1.1). The chosen single crystal was taken with a nylon loop and brought to the goniometer of the diffractometer together with the cooling device, maintaining in the cool nitrogen stream. In comparison to **12** $(Mes)_2PAlH_2·NMe_3$ appears to be more stable, since *Cowley* and *Jones* report a thermolysis reaction for this compound, that occurs at 95 °C.[61] The reason might be, that the trimethylsilyl-groups of **12** can be eliminated much more easily than the mesityl substituents in form of Me_3SiH. The crystal structure of **12** (Figure 15) possesses a mirror plane through the atoms P, Al, N and one methyl carbon atom of the amine base. This is the crystallographic mirror plane parallel to the *a*-axis of the space group $Cmc2_1$. The P–Al bond length (2.3428(15) Å) is shorter than in $(Mes)_2PAlH_2·NMe_3$ (2.409(3) Å) and in **4** (2.367(1) Å). In the first case the bulkier and less Lewis acidic mesityl groups are probably responsible for the elongated bond. In **12** the absence of the dimeric arrangement via Al–H⋯Al interactions in comparison to **4** provides an explanation for the slightly shorter bond. Since **4** shows a trigonal bipyramidal distortion of the coordination sphere at the aluminium atom the P–Al–N angle is widened to 106.97(9)° in **12** in comparison to this in **4** (103.28(8)°). The Si–P–Si' angle (107.02(5)°) is also widened compared to (Me$_3$Si)$_2$PAlMe$_2$·dmap (102.89(2), 104.82(2)°; two independent molecules within the asymmetric unit).[58] An explanation for this can be found in the arrangement of silyl and hydrogen substituents: it is *eclipsed* in the case of **12**, but the *gauche* conformer is reported for the comparable compound (Me$_3$Si)$_2$PAlMe$_2$·dmap. The larger methyl groups at the aluminium atom are probably responsible for this difference.

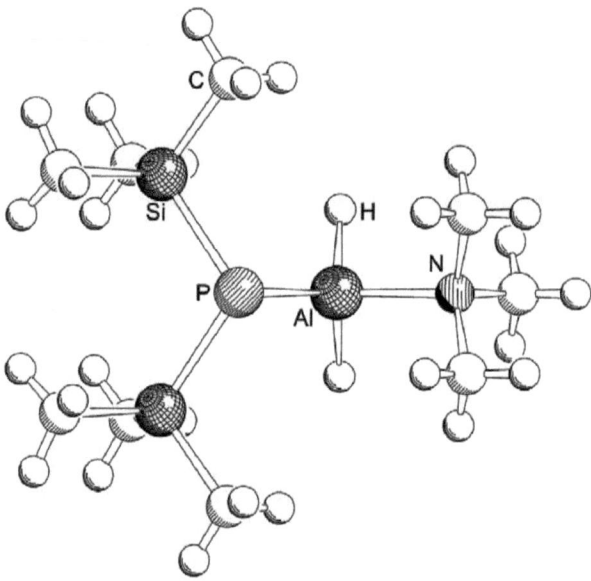

Figure 15: Molecular structure of **12** in the crystal. Selected bond lengths [Å] and angles [°]: P–Al 2.3428(15), P–Si 2.2378(9), Al–N 2.006(3), Si–P–Al 99.71(4), Si–P–Si' 107.02(5), P–Al–N 106.97(9).

During the efforts to isolate **12**, toluene was removed under reduced pressure. The following ^{31}P NMR spectroscopic investigations revealed a different major component at –275 ppm, besides the signal for **12**. The unit cell parameters of the obtained crystals together with the NMR data proved the trimerization product [(Me$_3$Si)$_2$PAlH$_2$]$_3$ which has been reported by *Wells* and *White* (Equation (19)).[64] A similar trimerization is also reported for (Mes)$_2$PAlH$_2$·NMe$_3$.[61] In this example the base gets eliminated upon heating the solid to

95 °C. If the reaction of (Me$_3$Si)$_2$PH and H$_3$Al·NMe$_3$ is carried out in dichloromethane, the solvent can be removed under reduced pressure and almost pure **12** can be isolated in yields of 50%. The volatile dichloromethane gets removed before the base elimination occurs.

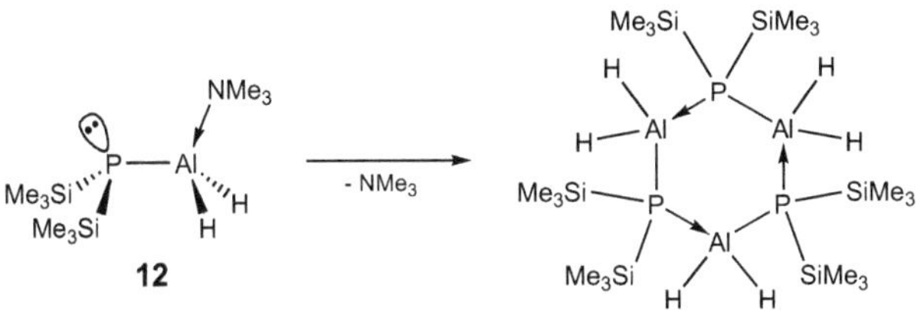

(19)

Theoretical Studies[52]

Theoretical computations were carried out on amine base eliminations (Scheme 11).[52] The synthesis of **12** is favourable (−23.7 kJ mol^{-1}). The base elimination reactions are very close in energy for a (hypothetical) dimerization (−1.8 kJ mol^{-1}) and the observed trimerization (+4.8 kJ mol^{-1}). Even a rearrangement of the dimer towards the trimer is only slightly endergonic with a Gibbs energy of +6.7 kJ mol^{-1}. However, no experimental evidence was found for the dimer.

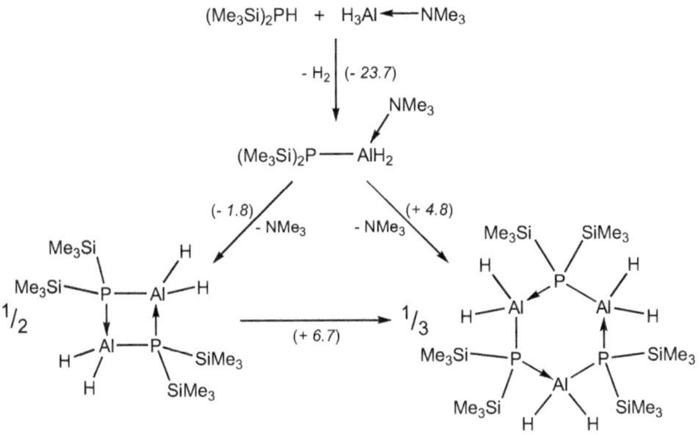

Scheme 11: Different reaction pathways to the formation of [(Me$_3$Si)$_2$PAlH$_2$]$_3$ (*Gibbs energies kJ mol^{-1}*).

Not taking into account the fact that the calculations are carried out for the gas phase, the experimental conditions differ from those used for the computation of the standard Gibbs energies. The amine base elimination has been observed under reduced pressure. During this treatment the solution gets cooled by the evaporation of the solvent. Hence, pressure and temperature are lower in the experiment compared to those used for the computation. As the calculated reaction energies are close to zero, an equilibrium is likely from thermodynamic data. However, the gaseous NMe$_3$ gets removed from the system. Thus, the equilibrium is pushed towards the trimerization product.

Compound **12** is the aluminium derivative of the intermediate for the synthesis of the LA-free H$_2$PBH$_2$·NMe$_3$ (Equation (11)). The reaction of a colourless solution of **12** with methanol immediately results in the formation of a white precipitate. The methanolysis of the silyl groups does not exclusively lead to a cleavage of the P–Si bond. Both, silicon and aluminium, possess a strong affinity towards oxygen. Hence, the desired reaction is in competition to the addition of methanol to aluminium. Furthermore, it has been shown, that CO$_2$ can even cleave a P–Al bond (Equation (8)).[27] This is also proven by the thermodynamic data of example reactions in Table 5.

Table 5: Thermodynamic characteristics for the gas phase reactions. B3LYP/6-31G* level of theory. $\Delta H°_{298}$ and $\Delta G°_{298}$ given in kJ mol^{-1}.

Process	$\Delta H°_{298}$	$\Delta G°_{298}$
(Me$_3$Si)$_2$PAlH$_2$·NMe$_3$ + MeOH = (Me$_3$Si)$_2$PAl(H)(OH)·NMe$_3$ + CH$_4$	−227.9	−219.8
(Me$_3$Si)$_2$PAlH$_2$·NMe$_3$ + MeOH = (Me$_3$Si)$_2$PAl(H)(OMe)·NMe$_3$ + H$_2$	−131.1	−116.6
(Me$_3$Si)$_2$PAlH$_2$·NMe$_3$ + MeOH = (Me$_3$Si)(H)PAlH$_2$·NMe$_3$ + CH$_3$OSiMe$_3$	−100.0	−107.9
(Me$_3$Si)$_2$PAlH$_2$·NMe$_3$ + 2 MeOH = H$_2$PAlH$_2$·NMe$_3$ + 2 CH$_3$OSiMe$_3$	−190.8	−200.8

Lewis-Acid-free Dimer

Since the methanolysis of the silyl groups in phosphanylalanes is inhibited by the competing reactivity with aluminium, other methods to obtain LA-free oligomers have to be employed. The previously described base elimination leads to a trimer, but contains dative bonds. Another possibility is to start from (Me$_3$Si)PH$_2$ instead of (Me$_3$Si)$_2$PH. The reaction with H$_3$Al·NMe$_3$ results in an elimination of two equivalents H$_2$. The expected dimer [(Me$_3$Si)PAlH·NMe$_3$]$_2$ (**13**) is formed and the amine base is still preventing dative bonding interactions.

$$(Me_3Si)PH_2 + H_3Al\leftarrow NMe_3 \xrightarrow{-2 H_2} \begin{array}{c} \text{Me}_3\text{Si}-\text{P}-\text{Al}-\text{H} \\ | \quad\quad | \\ \text{H}-\text{Al}-\text{P}-\text{SiMe}_3 \\ \text{Me}_3\text{N} \quad \mathbf{13} \end{array} \text{ with NMe}_3 \text{ on top P} \tag{20}$$

The reaction was carried out in dichloromethane at room temperature. During the solvent removal under reduced pressure, colourless blocks of **13** occur in the reaction mixture. The product is even less stable than **12**. White smoke is observed if the Schlenk tube is opened despite an argon stream and even at low temperatures. Furthermore, a white solid precipitates. For that reason the reaction was repeated in a NMR tube that had been dried at 200 °C under vacuum beforehand the experiment. The ^{1}H NMR spectrum of **13** shows two singlets, one for the trimethylsilyl hydrogens at 0.05 ppm and one for those of the amine at 3.32 ppm. A signal for the hydrogen atoms at the aluminium atoms is not observed. The phosphorus resonance is found at –284.5 ppm in the ^{31}P NMR spectrum, which is not split. This proves the absence of hydrogen substituents at the phosphorus atoms. The spectrum shows no additional signal for the monomeric intermediate. Another structural interpretation of the spectral data would be a trimer. However, as mentioned before six-membered ring motifs usually give different signals due to non-equivalent phosphorus and hydrogen atoms. Hence, the most likely description is a four-membered ring structure. Efforts to obtain a crytstal structure of **13** failed. The colourless blocks immediately turn white at the surface if the Schlenk tube is opened even at –30 °C. Those were anyway quickly brought to the diffractometer, but no reflections could be observed.

3.5. Introduction of a *N*-heterocyclic Carbene as Lewis Base

As strong Lewis-bases *N*-heterocylcic carbenes are interesting for stabilizing trielanes. In 1992 *Arduengo* reported the stable carbene 1,3,4,5-tetramethylimidazolin-2-ylidene (NHC^{Me}),[65] which was introduced as a LB at BH$_3$ by *Kuhn*[66]. Its aluminium derivative can be synthesized in diethylether at –50 to 0 °C (Equation (21)). After filtration, colourless needles of **14** are obtained at –28 °C. Removing the solvent under reduced pressure yields 50% of **14** as a white solid that slowly decomposes at room temperature. This is indicated by a colour change from white to grey. Storing the powder at –28 °C inhibits the decomposition.

$$3 \text{ LiAlH}_4 + \text{AlCl}_3 \xrightarrow[- \text{LiCl}]{\text{Et}_2\text{O}} 4 \text{ H}_3\text{Al} \leftarrow \text{OEt}_2 \xrightarrow[- \text{Et}_2\text{O}]{4 \text{ NHC}^{\text{Me}}} 4 \text{ H}_3\text{Al} \leftarrow \text{NHC}^{\text{Me}} \quad \textbf{14} \quad (21)$$

The ^{1}H NMR spectrum of **14** shows two singlets for the methyl groups at the nitrogen (1.14 ppm) and the carbon atoms (3.19 ppm). The broad signal at 4.5 ppm gives evidence for the aluminium-bound hydrogen atoms. In the ^{27}Al NMR spectrum the signal is detected at 106 ppm.

The X-ray structure analysis confirms the expected composition (Figure 16). The carbene carbon coordinates towards the alane with a bond length of 2.040(3) Å. In its chlorine-substituted derivative Cl$_3$Al·NHC^{Me} the corresponding dative bond is shortened to 2.009(5) Å by the electron withdrawing effect of chlorine atoms.[67] The C–N and C–C bond lengths of the carbene ligands are equal within the standard deviation range for both compounds.

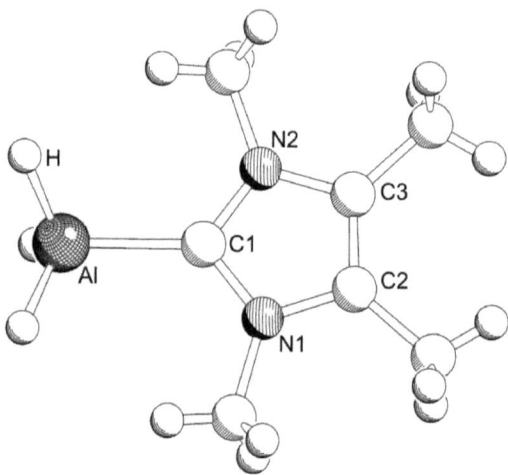

Figure 16: Molecular structure of **14** in the crystal. Selected bond lengths [Å]: Al–C1 2.040(3), C1–N1 1.354(3), C1–N2 1.356(3), N1–C2 1.387(3), N2–C3 1.390(3), C2–C3 1.347(4), Al–H 1.47 (restrained to equal bond lengths).

3.6. Stiba- and Bismaboranes

Very recently, the first hydrogen substituted, only LB-stabilized arsanylborane $H_2AsBH_2 \cdot NMe_3$ was synthesized in our group (Equation (11)).[25] Starting from this, efforts to introduce the heavier analogues antimony and bismuth were carried out.

Synthesis of $(Me_3Si)_2SbBH_2 \cdot NMe_3$ (15)

The reaction of $(Me_3Si)_2SbLi$ and $ClH_2B \cdot NMe_3$ at low temperatures in the dark leads to the formation of $(Me_3Si)_2SbBH_2 \cdot NMe_3$ (**15**) (Equation (22). The reaction mixture turns black during the synthesis. This fact indicates the presence of elemental antimony. However, after filtration of the crude reaction mixture, a pale yellow solution could be obtained, but turns black immediately if exposed to light. In a different approach the reaction of $(SiMe_3)_3Sb$ with $ClH_2B \cdot NMe_3$ does not form the desired product **15**.

$$(Me_3Si)_2SbLi + ClH_2B \leftarrow NMe_3 \xrightarrow{- LiCl} \underset{Me_3Si}{\overset{Me_3Si}{\diagdown}} Sb - BH_2 \quad NMe_3 \quad \mathbf{15} \tag{22}$$

The ^1H NMR spectrum of the filtrated reaction mixture of the reaction (22) is inconclusive due to closely neighbouring and overlapping signals for different $SiMe_3$ and NMe_3 groups and no B–H signals are observed. However, the ^{11}B NMR spectrum shows three boron containing compounds. Besides the triplet for the starting material $ClH_2B \cdot NMe_3$ and the quartet for $H_3B \cdot NMe_3$, a third triplet is detected at –8.5 ppm ($^1J(BH) = 114.8$ Hz) which can be assigned to **15**. This signal is broader than the others, giving evidence for a coupling to antimony (^{121}Sb: $I = 5/2$, ^{123}Sb: $I = 7/2$).

All efforts to obtain crystals of **15** failed to date, due to the instability of the compound. Even at –28 °C a darkening of the solution is observed after a few hours. This sensitivity is also displayed in the number of crystal structures of compounds with a direct Sb-B bond in the Cambridge Structural Database (CSD).[68] Only the LA/LB-adducts $X_3B \cdot Sb(SiMe_3)_3$ (X = Cl, Br, I)[69] and palladium-stibaborane *closo*-1,1-$(Me_2PPh)_2$-1,2,3-$PdSb_2B_9H_9$[70] are reported. A subsequent methanolysis of **15**, to obtain the LB-stabilized parent compound was carried out at different low temperatures (–30 to 0 °C) and with hexane/methanol and toluene/methanol mixtures. In all cases the solution turned black immediately. After filtration the ^{11}B NMR spectrum only shows signals for $ClH_2B \cdot NMe_3$ and the borane-amine adduct.

Bismuthaborane

The reaction of Bi(SiMe$_3$)$_3$ and ClH$_2$B·NMe$_3$ to obtain the bismuth analogue of **15** (Me$_3$Si)$_2$BiBH$_2$·NMe$_3$ results in a colourless solution with a black precipitate. The ^{11}B NMR spectrum of the filtrated crude reaction mixture only shows signals for ClH$_2$B·NMe$_3$ and H$_3$B·NMe$_3$.

Theoretical Aspects[52]

The thermodynamic data for the relevant reactions were computed to get insight into the observed reactivity of the compounds (Table 6). The LiCl eliminations (reactions *1*) are energetically favourable for antimony and bismuth. The reason that **15** can be synthesized, whereas its bismuth derivative does not form, might be found in the kinetic instability of the bismuth compound. Appearently it is less stable and even more sensitive towards light. The direct reaction with the (Me$_3$Si)$_3$E (E = Sb, Bi) is much less favourable than the LiCl elimination (reactions *2*). The methanolysis is energetically favoured (reactions *3*), but the resulting product is thermodynamically unstable towards the formation of the elemental metals and side products (example reactions *4*). Hence, the theoretical considerations match the experimental reactivity of the compounds. The instability of the (hypotetic) Bi–B bond is again proven by the CSD, in which no entries for compounds with a direct bismuth-boron-bond can be found.[63]

Table 6: Calculated thermodynamic characteristics, B3LYP/6-31G* (LANL2DZ(d) for Sb, Bi) level of theory. Data for the reactions including solid compounds are calculated using experimental values of sublimation enthalpies of 213.1, 268.2 and 209.2 kJ mol^{-1} for solid LiCl, Sb, and Bi, respectively, and sublimation entropies of 153.5, 134.5 and 130.0 J mol^{-1} K^{-1} for LiCl, Sb, Bi, respectively.

	Process	$\Delta H°_{298}$	$\Delta S°_{298}$	$\Delta G°_{298}$
1a	(SiMe$_3$)$_2$SbLi + ClBH$_2$·NMe$_3$ = (Me$_3$Si)$_2$SbBH$_2$·NMe$_3$ + LiCl (solid)	-222.8	-147.0	-178.9
1b	(SiMe$_3$)$_2$BiLi + ClBH$_2$·NMe$_3$ = (Me$_3$Si)$_2$BiBH$_2$·NMe$_3$ + LiCl (solid)	-208.3	-146.4	-164.7
2a	(SiMe$_3$)$_3$Sb + ClBH$_2$·NMe$_3$ = (Me$_3$Si)$_2$SbBH$_2$·NMe$_3$ + Me$_3$SiCl	-6.8	-4.3	-5.6
2b	(SiMe$_3$)$_3$Bi + ClBH$_2$·NMe$_3$ = (Me$_3$Si)$_2$SbBH$_2$·NMe$_3$ + Me$_3$SiCl	-15.0	-5.5	-13.3
3a	(Me$_3$Si)$_2$SbBH$_2$·NMe$_3$ + 2 MeOH = H$_2$SbBH$_2$·NMe$_3$ + 2 MeOSiMe$_3$	-202.7	26.0	-210.5
3b	(Me$_3$Si)$_2$BiBH$_2$·NMe$_3$ + 2 MeOH = H$_2$BiBH$_2$·NMe$_3$ + 2 MeOSiMe$_3$	-200.9	25.0	-208.4
4a	H$_2$SbBH$_2$·NMe$_3$ = Sb (solid) + H$_3$B·NMe$_3$ + $^1/_2$ H$_2$	-156.0	33.7	-166.1
4b	H$_2$BiBH$_2$·NMe$_3$ = Bi (solid) + H$_3$B·NMe$_3$ + $^1/_2$ H$_2$	-147.2	34.6	-157.5

4. Crystallographic Section

As a part of this thesis 132 crystal structures were determined. From those selected problematic examples are reported hereafter. The data for the other structures can be found in the attached CD.

4.1. General Procedures

4.1.1. Sample Handling

Most of the processed crystal samples were sensitive towards air and moisture. Some crystals turned out to be unstable at ambient temperatures. Thus, several techniques for sample preparation have been used.

Crystals were taken from a Schlenk flask under a stream of inert gas (argon, nitrogen) or directly from an evacuated, sealed ampoule inside a glove box and immediately put into a glass well containing a few drops of heavy mineral oil (Sigma Aldrich, CAS 8042-47-5) or perfluorinated hydrocarbons (Fomblin, Aldrich, CAS 69991-67-9). Low temperature samples were taken from a cooled Schlenk tube and brought into a stream of cold nitrogen (ca. 170–200K). Those were either prepared directly or put into small sample carriers containing perfluorinated polyethers (Galden, Solvay Solexis S.p.A, CAS 69991-67-9) After investigation using an optical microscope the chosen crystal, together with some oil, was transferred to a glass fibre or a CryoLoop ($\varnothing$ 0.1-1.0 mm, 20 µm, Hampton Research) of a goniometer head, subsequently brought into the cold nitrogen stream of the Oxford Instruments CryoJet (Oxford Diffraction Gemini R Ultra devices) and Oxford Cryosystems Cryostream 600 (Oxford Diffraction SuperNova), respectively.

4.1.2. Data Collection

The data were acquired either at an Oxford Diffraction Gemini R Ultra diffractometer using Cu or Mo radiation from sealed tubes and a Ruby CCD detector, or at an Oxford Diffraction SuperNova device employing a microfocus copper source with Atlas CCD detector. The experiment 'boden_5_175' was carried out in Prague in collaboration with *M. Dušek* at an Oxford Diffraction Gemini R Ultra using Mo radiation from a sealed tube and an Atlas CCD detector. After crystal centring, usually 30 frames in six different goniometer orientations and four different detector angles were collected to determine the unit cell from independent reflections and the orientation matrix of the crystal using CrysAlis software.[71] Parameters like indexation, unit cell standard deviations and mosaicity were checked to estimate the quality of the crystal. In particular, the 'G^6 projection distance' value[72] was controlled for consistency with the Laue group suggested by the software. Finally the software strategy tool was used to compute an experiment with optimal coverage, resolution, ω-increment and average $I/\sigma(I)$.

4.1.3. Data Processing

After the data collection the unit cell parameters as well as the orientation matrix were refined using the reflections of the whole experiment. Subsequently, the intensities were determined by integration using '3D profile fitting' and 'smart background' subtraction to gain the raw intensity data file. In special cases (e.g. twinning) an 'average background' subtraction had to be employed. Furthermore, Lorentz and polarization corrections were automatically applied by the software. Finally, a semi-empirical multi-scan absorption correction from equivalents was applied to generate the hkl-file for the structure solution. If necessary, an analytical absorption correction from crystal faces was applied after the absorption coefficient (μ) had been determined from elemental analysis, expected chemical composition or the final structure model.[73]

4.1.4. Space Group Determination

Except for 'boden_5_175' (details in chapter 4.6), the space groups were determined using the programs GRAL[71] and XPREP[74], respectively, starting from the metric parameters of the unit cell together with the Laue group found by the 'G^6 projection distance' value. After verifying Bravais lattice exceptions, higher metric symmetry was checked using the reduced cell. The space groups were determined by systematic absence conditions together with the $|E^2-1|$ value to estimate the presence of an inversion centre. Those E-values are normalized to θ-independent structure factors for several statistic evaluations and structure solutions applying direct methods. One of these evaluations is the comparison between the theoretically computed $|E^2-1|$ for centrosymmetric (0.968) and non-centrosymmetric (0.736) unit cells, assuming a statistical distribution of atoms with the same electron number.[75] However, values closer to one are widely used to estimate the presence or absence of a crystallographic inversion centre. Additionally, a lowered $|E^2-1|$ often indicates twinning (Chapter 4.3). For

4. CRYSTALLOGRAPHIC SECTION

the final space group decision in XPREP the combined figure of merit (CFOM) was counterchecked. The CFOM is a quality factor gained from the combination of the previously mentioned indicating data, and a lower number means a better agreement.

4.1.5. Structure Solution and Refinement[47]

The target of the following structure solution is to create a three dimensional electron density map (ρ_{xyz}) within the volume of the unit cell (V) from these data. Density maxima can be assigned to the corresponding atoms by their electron number. Mathematically the density map can be calculated by a Fourier synthesis of the structure factors (F_{hkl}).

$$\rho_{xyz} = \frac{1}{V} \sum_{hkl} F_{hkl} \cdot e^{-i2\pi(hx+ky+lz)}$$

These structure factors sum up from the atomic form factors (f), which are tabulated functions of theta angle and wavelength. The obtained intensities from the experiment are proportional to the squared structure factors but the phase information is lost. Thus statistical methods were developed to determine them from the intensities.

Direct methods compute normalized structure factors (E) independent on the diffraction angle (θ) to find strong reflections with most likely correct phases. In a final Fourier synthesis from the best solution the starting model of the structure is obtained.

The Patterson or heavy atom methods directly employ the intensities to receive the difference vectors between atoms. Since phase information is absent only their relative positions are obtained. The Patterson maxima are proportional to the product of the atomic numbers of two atoms connected by such a vector. For this reason the method only works if larger differences in the atomic numbers between the involved elements are given. It works best with a single or few heavy atoms carrying a large amount of electrons of the cell. The absolute coordinates are gained applying the symmetry of the space group.

Charge-flipping methods[76] have been developed very recently. If necessary, the input reflections are extended to those of the whole unit cell applying symmetry equivalency rules. Subsequently, an arbitrary set of starting phases is assigned to compute a first electron density map. Since this density has to be positive, values below an empirically estimated slightly positive δ value are inverted and new structure factors are calculated from it. Together with the data from the experiment the next cycle of the iteration is started. The final electron density map can be utilized to check the correctness of the space group, since the density is refined for the whole unit cell. In the last step this map has to be interpreted to assign the atoms.

Different programs implemented in WinGX[77] have been used for solving the structures of this thesis. Except for the structure 'boden_5_175' (details in chapter 4.6), SHELXS-97[78] was used for solutions with direct or Patterson methods. SIR[79] can also apply those methods, but for the following structures only direct methods were employed if SIR was used. For structures solved by charge-flipping methods Superflip[80] (density map generation) has been used together with EDMA[81] (interpretation).

Starting from the first structure solution further refinement is required to achieve the optimized final model. In most of the cases several atoms are missing and additional effects like atomic vibration, disorder, twinning and other factors have to be considered. Generally the least-squares method on F^2 was applied to optimize all parameters. Difference Fourier syntheses were employed to locate residual density maxima. Here only the basic procedures for all structures are mentioned. Beyond that, special refinement features are discussed in the individual chapters.

Oscillation of atoms was taken into account by modifying the atomic form factors with additional terms for isotropic (hydrogen atoms, minor disordered parts) or anisotropic displacements (U).

$$f_{iso} = f \cdot \exp\left(-8\pi^2 U \frac{\sin^2\theta}{\lambda^2}\right)$$

$$f_{anis} = f \cdot e^{-2\pi^2(U^{11}h^2a^{*2}+U^{22}k^2b^{*2}+U^{33}l^2c^{*2}+2U^{23}klb^*c^*+2U^{13}hla^*c^*+2U^{12}hka^*b^*)}$$

To be able to evaluate the accordance of the model's computed structure factors (F_c) and intensities (F_c^2) with the measured ones (F_0, F_0^2), different quality factors are used. R_1 gives the deviation of the structure factors,

$$R_1 = \frac{\sum_{hkl}||F_0|-|F_c||}{\sum_{hkl}|F_0|}$$

whereas wR_2 compares the intensities and further contains a weighting factor. This considers experimental standard deviations and possible systematic errors as two refined variables

4. CRYSTALLOGRAPHIC SECTION

(a, b). Due to the squared values and the weighting scheme, this value is more sensitive than R_1.

$$wR_2 = \sqrt{\frac{\sum_{hkl} w(F_0^2 - F_c^2)^2}{\sum_{hkl} w(F_0^2)^2}}$$

The goodness-of-fit takes the weighting as well as the data (m) to parameter (n) ratio into account and thus also indicates the redundancy of the data.

$$S = \sqrt{\frac{\sum_{hkl} w(F_0^2 - F_c^2)^2}{m - n}}$$

Besides 'boden_5_175' (chapter 4.6), all structures have been refined using the least squares method on F^2 employing SHELXL-97[78] and given quality factors are computed for $F^2 > 2\sigma(F^2)$. During the structure refinement SXGraph[77] and Ortep3[82] were used to depict the models. Figures were created with Schakal99[83], Diamond 3[84] and Topos40[85].

4.2. Disorder in Mixed Crystals

Disorder is a common problem in crystallography, which can make the least-square refinement challenging, due to closely neighbouring or overlapping disordered positions and correlations between them. Consequently, constraints for the atomic positions and restraints for interatomic distances and anisotropic displacement parameters are often required.[86] The final model shows an asymmetric unit containing all possible atomic positions and their relative amounts. Hence, it is initially impossible to derive their distribution in the different unit cells of the crystal. It can either be statistical or systematic. Only if positions cannot exist together or are related by crystallographic symmetry elements conclusions can be drawn.

From the reaction mixture of the phosphinidene complex [Cp*P{W(CO)$_5$}$_2$] reduced by [CoCp$_2$] four differently shaped crystals could be obtained. Not taking crystals of the side product [W(CO)$_5$Cl][CoCp$_2$] (ms247b) into account, the remaining three consist of two or three of the following compounds together with the [CoCp$_2$]$^+$ counterion in mixed crystals (Figure 17). Depending on the chemical composition, they even show different crystal systems. Several experiments were carried out to investigate this system (Table 7).

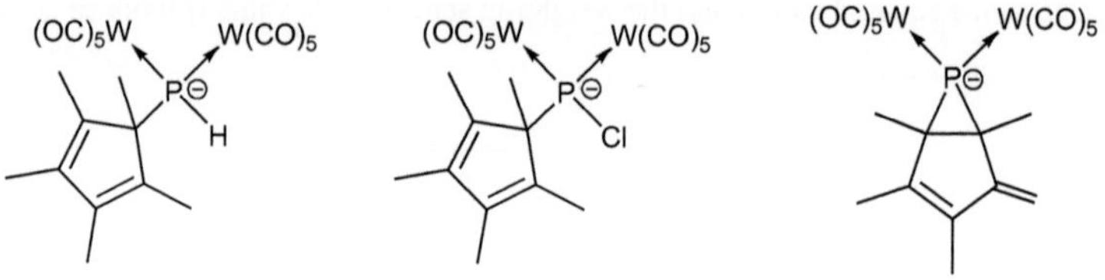

Figure 17: Different anionic compounds occurring in the mixed crystals. [CoCp$_2$]$^+$ counterion not depicted.

Table 7: Data overview of the processed crystals of ms247.

	ms247	ms247a	ms247c	ms247d	ms247e	ms247f	ms247g
a [Å]	35.9341(5)	9.8971(4)	9.1967(1)	35.8694(4)	9.8979(4)	9.2018(2)	35.8910(8)
b [Å]	35.9341(5)	10.8277(4)	30.2845(3)	35.8694(4)	10.8361(4)	30.2736(8)	35.8910(8)
c [Å]	13.2128(3)	16.0942(5)	11.6826(1)	13.2605(2)	16.0695(7)	11.6671(3)	13.2467(4)
α [°]	90	89.217(3)	90	90	89.056(3)	90	90
β [°]	90	75.095(3)	98.012(1)	90	75.227(4)	97.966(2)	90
γ [°]	120	78.992(3)	90	120	78.868(3)	90	120
V [Å^3]	14775.4(4)	1634.86(11)	3222.05(6)	14775.4(3)	1634.22(12)	3218.76(14)	14777.8(6)
crystal system	trigonal	triclinic	monoclinic	trigonal	triclinic	monoclinic	trigonal
space group	$R3$	$P\overline{1}$	$P2_1/c$	$R3$	$P\overline{1}$	$P2_1/c$	$R3$
data/unique (R_{int})	11851/6325 (0.0255)	11329/6329 (0.0282)	40923/6429 (0.0398)	20055/6478 (0.0414)	11350/6384 (0.0297)	12747/6295 (0.0342)	21930/6461 (0.0343)
parameters	361	412	465	446	414	461	522
restraints	62	13	4	121	1	4	2
R_1	0.0444	0.0394	0.0330	0.0381	0.0320	0.0382	0.0344
wR_2	0.1066	0.1005	0.0747	0.0879	0.0818	0.0891	0.0796
S	1.066	1.100	1.143	1.025	1.076	1.049	1.028
ratio **A:B:C**	79:21:0	43:57:0	25:9:66	72:28:0	43:57:0	28:8:64	77:23:0

The crystals featuring the triclinic crystal system possess a chlorine position, which is not fully occupied compared to the connected phosphorus atom. Hence, the following least-square refinement cycles were performed with alternating fixed isotropic displacement parameter and occupancy, respectively. The final model shows compounds **A** and **B**. Hereby, only the chlorine and hydrogen positions of the molecule are disordered.

In contrast, the structure solutions of the trigonal sample already reveal two disordered phosphorus positions. Consequently, the Cp* arrangement is also affected. Furthermore, the [CoCp$_2$]$^+$ counterion is disordered. A residual electron density peak in a distance of about 2 Å to the second phosphorus atom can be assigned to a chlorine position. Interestingly, it is again less occupied than the phosphorus atom it is attached to. Thus, its isotropic displacement parameter and occupancy were fixed and refined in an alternating manner. That way treated, a model results in a 21 to 28 per cent occupied chlorine position, depending on the processed crystal. Hence, the second phosphorus atom has to carry a hydrogen substituent in the other case. The hydrogen positions were refined with fixed P–H distances. The carbon atoms in the minor component were refined isotropically, and several restraints were applied to avoid correlation effects. However, again **A** and **B** are found, but **A** is disordered over two positions, of which one is a mixed PH/PCl position (Figure 18).

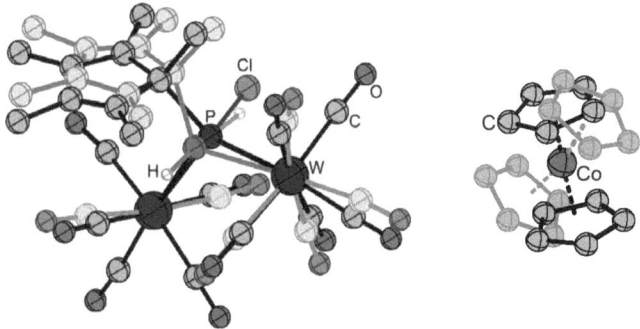

Figure 18: Disorder of both ions in the trigonal crystals. Main parts depicted in black, minor parts in grey.

A third component can be found in the case of the monoclinic crystals. Again the phosphorus position is split. The major one belongs to the anion **C** and the minor one is a mixed position of **A** and **B** (Figure 19). In contrast to the trigonal space group, the $[CoCp_2]^+$ is not disordered. However, Cp* is slightly disordered, but at least the ring carbons cannot be refined separately. This leads to an apparently wrong geometry within the five-membered ring, due to differently located double bonds in **C** compared to **A** and **B**. However, the exocyclic double bond is found disordered over two positions. For these disorder reasons, the carbon atoms of the minor component and the chlorine atom were kept isotropic and several restraints had to be applied.

A trend can be derived from the observed space groups and the composition of the crystals. Compound **C** is only found in the monoclinic crystals, and its packing motif allows minor amounts of **A** and **B** to be included. If the crystal only consists of **A** and **B**, the space group depends on their relative amounts. A predominant **A** forces the trigonal one, whereas an almost equal ratio results in the triclinic crystal system.

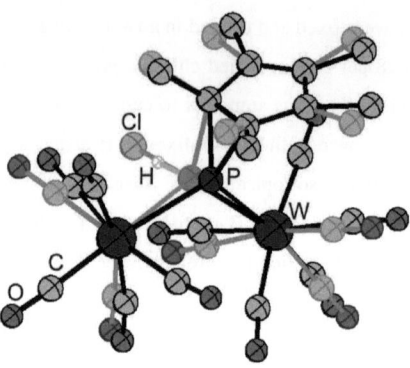

Figure 19: Disorder of the anionic part of the monoclinic crystals. Main part depicted in black, minor parts in grey.

4.3. Twinning[87]

Twinning is a crucial problem in crystal structure determinations since the method strictly requires a single crystal. For the description of twins it is necessary to know the relative orientations of the twin components ('twin law') and the amounts of both. Four different types are possible. Twins by merohedry are not directly visible from the diffraction pattern, since the reciprocal lattices of the domains superimpose each other. This is a consequence of the twin law being a symmetry operator of the crystal system, but not of the point group of the crystal. Two subtypes are possible. Firstly, the twin law belongs to the same Laue group as a simple inversion (racemic twinning). Secondly, the twin operator does not belong to the same Laue group, but to the same crystal system. Hence, it is only possible in the trigonal, hexagonal, tetragonal and cubic crystal systems. Pseudo-merohedral twinning requires the metric symmetry of the unit cell to be higher than the symmetry of the space group. Again, the twinning is not obvious from the diffraction pattern. In the case of reticular merohedral twinning not all reflections are affected. Systematic absence conditions of one component are violated by the other. Finally non-merohedral twinning is visible in the diffraction pattern. Hereby, the orientation of the domains is arbitrary and usually the majority of the reflections do not superimpose each other.

Twinning is often indicated by certain observations. For instance non-merohedral twinning can already be seen in the diffraction pattern, since reflections appear to be split. (Pseudo)-merohedral twins often show a significantly lowered $|E^2-1|$ value. Other indicators are e.g. inconsistent systematic absences, similar R_{int}-values for different crystal systems, a metric symmetry higher than the Laue symmetry and generally no structure solution or bad quality factors from apparently good data.

4.3.1. Pseudo-merohedral Twinning

The experiment 'hfk140' gave evidence for twinning. From metric point of view the unit cell appeared to be orthorhombic with all angles very close to 90°, but a look into the 'G^6 projection distance' values shows a remarkably low number for one of the possible monoclinic Laue classes (Table 8)

Table 8: Lattice transformation results from CrysAlis 'Lattice Wizard'. G^6 projection distance is zero for the triclinic cell per definition.

#	IT Code	transformed cell (a,b,c,al,be,ga,vol)							G6 proj dist
1	32 oP	8.47860	11.67896	21.67304	89.99328	89.96952	89.99148	2146.09	0.10321
2	33 mP	8.47860	11.67896	21.67304	90.00672	90.03048	89.99148	2146.09	0.03315
3	34 mP	8.47860	21.67304	11.67896	89.99328	90.00852	90.03048	2146.09	0.10215
4	35 mP	11.67896	8.47860	21.67304	90.03048	90.00672	89.99148	2146.09	0.09884
5	31 aP	8.47860	11.67896	21.67304	89.99328	89.96952	89.99148	2146.09	0.00000

Hence, this setting was chosen and the space group $P2_1/c$ was used as the only suggestion of XPREP. Interestingly the $|E^2-1|$ value is 0.772 indicating a non-centrosymmetric space group. As mentioned before, such a lowering can point towards twinning.

Superflip was used for structure solution to countercheck the space group derived from electron density and also results in $P2_1/c$. All heavy atoms are found and after some least-square refinement cycles the carbon atoms could also be assigned, but the refinement is unstable. Additionally, the geometry of the molecule, especially that of the Cp*-ligands seems distorted and its C–C-bond lengths within the five-membered ring differ severely, and show unusually high standard deviations (1.22(5)–1.64(10) Å). Furthermore, high residual electron density occurs at meaningless positions, quality factors are significantly high as well as the second parameter of the weighting scheme b (Table 9) and almost all C and N atoms are fitted by non-positive definite anisotropic displacement parameters. The sum of these facts points towards pseudo-merohedral twinning in which the twin operator and the metric symmetry belong to a higher crystal system. PLATON's 'TwinRotMat' function in fact detects a two-fold axis along a as twin operator (a, –b, –c) and estimates the batch scale factor to 0.40.[88] Employing the adequate twin matrix for further least-square refinement gives much better results and the batch scale factor of the twin component refines to 0.422(1).

Table 9: Comparison between the two different models.

	single crystal model	twin model
R_1	0.267	0.0427
wR_2	0.587	0.1015
S	1.147	1.042
b	2222.33	-
$\Delta\rho$	7.45/–5.12 eÅ^{-3}	1.74/–1.41 eÅ^{-3}

The twin model's residual electron density is located close to the heavy atoms and the geometry of the molecule is no longer distorted: The bond lengths within the Cp*-ligand show normal values (1.416(8)–1.434(8) Å) and all atoms are fitted by positive definite anisotropic displacement parameters with similar values.

4.3.2. Merohedral in Combination with pseudo-merohedral Twinning

The raw data unit cell angles of 'em144_1' indicate the orthorhombic crystal system (90.02, 90.04, 89.97). The space group determination procedure shows no systematic absence exceptions besides the C-centring, and $|E^2-1|$ is inconclusive (0.882) (Table 10).

Table 10: Space group suggestions of XPREP for 'em144_1' in the orthorhombic crystal system.

```
Option  Space Group  No.    Type      Axes  CSD  R(sym)  N(eq)  Syst. Abs.   CFOM
[A]     C222         # 21   chiral    1     19   0.027   8560   0.0 / 5.9    8.82
[B]     Cmmm         # 65   centro    1     7    0.027   8560   0.0 / 5.9    14.92
[C]     Cmm2         # 35   non-cen   1     1    0.027   8560   0.0 / 5.9    53.82
```

Hence, the space group with the lowest symmetry $C222$ was chosen for the first structure solution employing SIR. However, the solution appears wrong due to a strongly distorted geometry and very high R-values ($R_1 > 0.3$, $wR_2 > 0.7$) in the following refinement, even if inversion twinning is considered. Obviously, a lowering of the Laue class to monoclinic C-centred was indicated, taking pseudo-merohedral twinning into account. In the subsequent space group determination XPREP suggests three different monoclinic space groups, all of which showing significantly lowered combined figures of merit compared to the orthorhombic space groups (Table 11).

4. CRYSTALLOGRAPHIC SECTION

Table 11: Space group suggestions of XPREP for 'em144_1' in the monoclinic crystal system.

Option	Space Group	No.	Type	Axes	CSD	R(sym)	N(eq)	Syst. Abs.	CFOM
[A]	C2	# 5	chiral	1	552	0.022	5106	0.0 / 7.1	3.72
[B]	C2/m	# 12	centro	1	310	0.022	5106	0.0 / 7.1	2.47
[C]	Cm	# 8	non-cen	1	32	0.022	5106	0.0 / 7.1	6.57

$C2/m$ was chosen due to the best combine figure of merit. A subsequent structure solution and refinement gave wrongly connected molecules, residual density peaks in obviously wrong positions and the R-values remained high ($R_1 > 0.2$, $wR_2 > 0.5$). The following structure solution used Superflip to obtain the space group from electron density and resulted in $C2$. Hence, the least-square refinement was carried out in this space group and pseudo-merohedral twinning ($-a, -b, c$) was applied to the model. It results in good quality factors ($R_1 = 0.0653$, $wR_2 = 0.1707$, $S = 1.086$), but the refinement does not converge and the Flack parameter refines to 0.43(2). Hence, additional racemic twinning was taken into account, which results in a final model with better quality factors ($R_1 = 0.0594$, $wR_2 = 0.1534$, $S = 1.040$). The amounts of the twin components are refined to $k_2 = 0.201(4)$ (–1 0 0, 0 –1 0, 0 0 1), $k_3 = 0.203(16)$ (–1 0 0, 0 –1 0, 0 0 –1) and $k_4 = 0.299(5)$ (1 0 0, 0 1 0, 0 0 –1). The better combined figure of merit for the space group $C2/m$ compared to the real one $C2$ is caused by the additional racemic twinning. As a consequence of the pseudo-merohedral twinning the strategy for the experiment was erroneously computed for the orthorhombic Laue class, resulting in a completeness of only 0.89. In addition to this several restraints were applied to avoid pseudo-symmetrical correlations and to model co-crystallized disordered toluene molecules.

4.4. Space Group Problem

The refinements of the chemically related compounds $Cl_4E \cdot NHC^{dipp} \cdot 2(C_7H_8)$ with E = Sn ('ED1') and Ge ('ED3', 'ED3a') are interesting from a crystallographic point of view. Both compounds possess similar reduced unit cell parameters (ED1: $a = 14.6292(5)$, $b = 16.2769(6)$, $c = 17.3950(6)$ Å; ED3a: $a = 14.53580(1)$, $b = 16.0001(1)$, $c = 17.5332(1)$ Å). For the tin compound the space group is clear, but in the case of its germanium derivative the space group decision is less obvious. Here, already the sample preparation is problematic. Due to the much higher sensitivity of the germanium compund compared to its tin derivative, the crystals decompose very quickly beginning from the surface if put into any kind of oil, and crack if treated in a excessively cold N_2-stream. Thus the crystals were taken as fast as possible from the Schlenk tube to mineral oil and then to the goniometer. However the crystals edges and vertices became smooth during this procedure and the surface turned white.

For the tin derivative the space group determination using XPREP only suggests the chiral space group $P2_12_12_1$. This is indicated by strong systematic absence violations for any of the possible glide planes in the orthorhombic crystal system (Table 12).

Table 12: Systematic absence exceptions report and space group suggestion of XPREP for 'ED1'.

```
             b--    c--    n--    21--   -c-    -a-    -n-    -21-   --a    --b    --n    --21
N            249    252    247    9      236    238    236    10     207    212    211    11
N I>3s       176    178    162    0      146    140    92     0      118    93     125    0
<I>          157.2  128.1  130.7  1.6    115.0  115.6  19.0   3.0    124.2  24.4   130.1  1.2
<I/s>        10.0   9.0    8.8    0.5    8.1    8.3    3.4    0.7    7.3    3.9    7.9    0.3

Option   Space Group      No.   Type     Axes   CSD    R(sym)  N(eq)   Syst. Abs.    CFOM
[A]      P2(1)2(1)2(1)    19    chiral   1      5917   0.047   4107    0.7 / 3.4     3.31
```

Further the $|E^2-1|$ value is at 0.794 very close to the theoretical value for non-centrosymmetric space group. For the model in $P2_12_12_1$ the Flack parameter[89] was found to be 0.266(12), and thus, racemic twinning was used for further refinement. The final quality factors are within the range of a good structure solution ($R_1 = 0.046$, $wR_2 = 0.097$, $S = 1.030$). Interestingly, PLATON[88] detects probable additional symmetry towards *Pnma*. This can be an effect of a different threshold used by PLATON compared to XPREP. Refinement in this space group results in much worse quality factors ($R_1 = 0.104$, $wR_2 = 0.236$, $S = 1.346$). Additionally one of the solvent molecules becomes disordered in the centric model, indicating that the higher symmetry is only violated by the toluene molecule. Despite the large correlations between the pseudo-symmetrically related carbon and nitrogen atoms of the non-solvent parts of the structure, no correlations for the toluene molecule are reported in the listing file of the acentric refinement, which further supports this assumption. Further the systematic absence report gives $I/\sigma(I)$ values of 3.4 and higher for any of the glide plane extinctions. Those can be caused by only few atoms in violating positions compared to a larger amount of other atoms (especially heavier ones) resulting in a pseudo-symmetry. Their contribution to the structure factors is small, but become visible in violating systematic absence conditions (Figure 20).

4. CRYSTALLOGRAPHIC SECTION

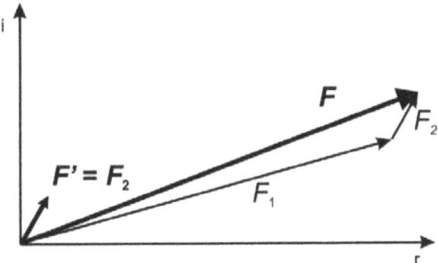

Figure 20: Structure factors for non-absent (F) and absent reflections (F'): Contribution of few symmetry violating atoms (F_2) to the pseudo-symmetrical parts of the structure (F_1).

In extreme cases a single atom can even force a change of the crystal system. For example, the only chiral methyl group in (+)–camphoric acid as the ligand in a coordination polymer reduced the symmetry from the acentric tetragonal space group $I4c2$ to the chiral orthorhombic space group $F222$, undergoing a maximal non-isomorphic *translationsgleiche* transition.[90] As reported by Flack, PLATON can erroneously claim a structure to be centrosymmetric, but the Flack parameter clearly indicates this failure.[91]

The first experiment for 'ED3' at the Gemini Ultra R diffractometer could only provide data to a resolution of 0.93 Å within reasonable time, but already shows difficulties in finding the proper space group. This problem could only be investigated in detail after the SuperNova system had been installed, owing to a much stronger microfocus source. Employing this opportunity a better experiment ('ED3a') could be carried out to a resolution of 0.79 Å. XPREP only suggests *Pnma* and *Pna*2_1 space groups. The first is chosen by the software due to the $|E^2-1|$ value of 0.954 together with only weak systematic absence violations. However, also the subgroup $P2_12_12_1$ has also to be taken into account for the following discussion (Table 13).[92]

Table 13: Systematic absence exceptions report and space group suggestions (automatic decision tolerances changed to display $P2_12_12_1$) of XPREP for 'ED3a'.

	b--	c--	n--	21--	-c-	-a-	-n-	-21-	--a	--b	--n	--21
N	1420	1425	1423	39	1177	1184	**1175**	46	1287	**1292**	1279	43
N I>3s	1193	1191	1182	1	607	623	**180**	1	626	**218**	652	1
<I>	23.5	17.3	18.4	0.1	13.8	13.8	**0.2**	0.1	19.2	**0.3**	19.4	0.1
<I/s>	25.9	22.9	22.0	0.7	14.4	14.4	**1.7**	0.8	15.0	**1.8**	15.2	0.9

Option	Space Group	No.	Type	Axes	CSD	R(sym)	N(eq)	Syst. Abs.	CFOM
[A]	P2(1)2(1)2(1)	# 19	chiral	1	5917	0.015	10779	0.9 / 1.7	14.13
[B]	Pna2(1)	# 33	non-cen	3	903	0.015	10779	1.8 / 14.4	5.64
[C]	Pnma	# 62	centro	2	894	0.015	10779	1.8 / 14.4	0.93

Refinement in this space group gives a good model for the main part of the structure. A toluene molecule disordered over two positions with 50 percent occupancy can be located if isotropic displacement parameters are kept. For the second, more severely disordered solvent molecule the occupancies were refined for all carbon atoms in an isotropic model resulting in a well converged model with 50 percent disordered carbons, but proper geometry could not be modelled. Lowering symmetry to the *klassengleiche* non-centrosymmetric subgroup $Pna2_1$ does not improve the disorder situation. In $P2_12_12_1$, the chiral space group of its Sn derivative, the toluene molecule is still disordered. Interestingly, the changes to the acentric space groups lead to significantly better quality factors, but also causes large correlations between the pseudo-centric parts of the structure (Table 14).

Table 14: Comparison of the refinement in different space groups.

space group	*Pnma*	*Pna2₁*	$P2_12_12_1$ (split Cl1)
R_1	0.0621	0.0478	0.0497 (0.0475)
wR_2	0.1646	0.1294	0.1309 (0.1289)
S	1.080	1.047	1.024 (1.022)
Flack paramter	-	0.49(3)	0.49(3) (0.51(3))
data/parameter ratio	4427/233	7887/416	8461/404 (414)

The $P2_12_12_1$ model shows a large anisotropic displacement for Cl1 and was hence refined at two split positions. Despite the higher symmetrical space group, $Pna2_1$ is favourable from point of view of the quality factor, the systematic absence violations still point towards $P2_12_12_1$.

As mentioned before, a few atoms can already force a change of the space group. However, another explanation for systematic absence violations is often found in the quality of the processed crystal. For example, the presence of small additional single crystals is overlooked or the crystal decomposes to powder, losing solvent. Both can result in observed intensity at unexpected positions. Since decomposition was clearly visible for this toluene-containing sample, and the wrong intensity can be seen in the low angle frames, this is the most probable explanation for the systematic absence violations (Figure 21). Hence, $Pna2_1$ is the most likely space group for 'ED3(a)'.

Theoretically, a different crystal could be a single one or racemically twinned in a ratio far from 50:50 as shown in ED1. This could finally prove the $P2_12_12_1$ space group, but all efforts to find such a crystal failed. However, it is very rare that such closely related compounds with almost equal cell parameters are non-isostructural. Hence, the true chiral space group $P2_12_12_1$ may be hidden by twinning, but $Pna2_1$ provides the best model.

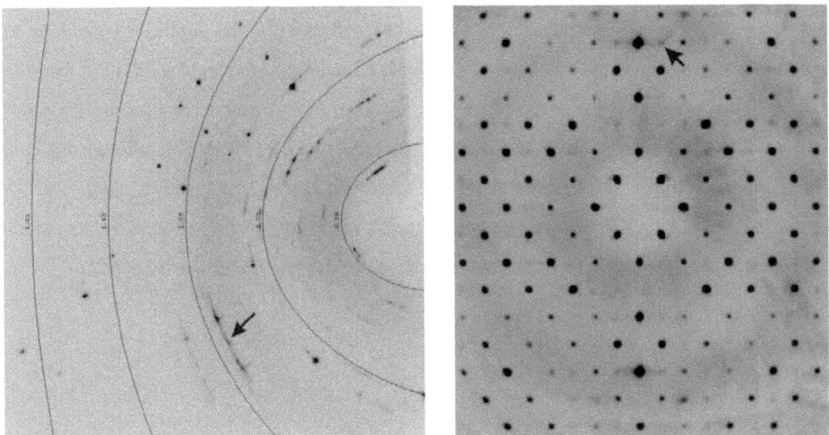

Figure 21: Decomposition of the crystal 'ED3a' visible in the low angle frames (left); systematic absence violations for n glide plane ($k + l \neq 2n$) in the 0kl plane (right). Arrows mark the actually extinguished example reflection 0 2 7.

4.5. Disordered Solvent Treatment Applying SQUEEZE[93]

The SQUEEZE function implemented in the PLATON software is a tool to handle severely disordered solvent molecules. The contribution of the solvent to the total structure factor is subtracted by a discrete Fourier transformation and incorporated in a further least-squares refinement of the ordered part. This results in complete exclusion of whole molecules, allowing good structure solutions. Furthermore, SQUEEZE outputs an overview of the size and volume of the corresponding void(s) along with the number of electrons it/they contain(s). However, this powerful tool can be abused to omit parts of a structure that could be refined with certain effort or even lead to bigger errors. For instance, applying SQUEEZE in the case of 'ED3a' in the previous chapter, the severely disordered toluene molecule could be removed, resulting in the wrong space group *Pnma*. Nevertheless it can help to determine the solvent and thus clarify the chemical composition, as shown in the case of the polymer $[Cp*_2Mo_2P_2Se_3(P_4Se_3)(CuI)_2]_n$ ('pmmv33n').

During the refinement of the structure the P_2Se_3 middle deck as well as the P_4Se_3 cages are found to be disordered with 50 per cent occupancy for each site (R_1 = 0.056, wR_2 = 0.174, S = 1.093). However, large residual electron densities are located within a huge channel-shaped void. The chemist had used a mixture of solvents for the synthesis, in which an acetonitrile solution of copper iodide was layered over a dichloromethane solution of the other starting compound. The two highest peaks in the asymmetric unit were thus assigned to a semi-occupied chlorine position and a quarter-occupied one. The model refines well even employing anisotropic displacements for the chlorine atoms (R_1 = 0.043, wR_2 = 0.125, S = 1.018). A view on the solvent part of the structure, including symmetrically related atoms and peaks, gives evidence for additional, disordered acetonitrile molecules (Figure 22).

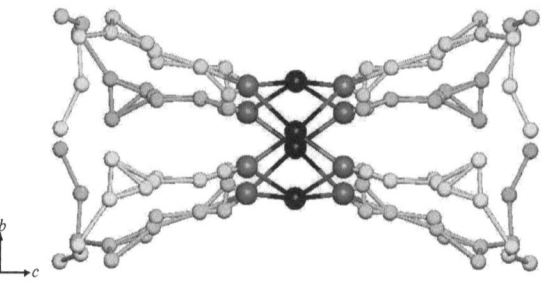

Figure 22: View to the solvent part of the structure along the a-axis ($0 < c < 1$). Semi-occupied (black) and quarter-occupied (dark grey) chlorine positions. Residual density peaks (0.60 to 1.84 electrons, light grey).

Since already the assigned chlorine positions have to be mixed Cl/C positions to form reasonable dichloromethane molecules and the acetonitrile molecules appear to be even more disordered, SQUEEZE was applied. The report scales the size of the void to 938 Å^3 and sums up the electron number within it to 170. These data match two molecules of CH_2Cl_2 and four acetonitrile molecules (172 electrons) per void. Refinement on the hkl-file results in good quality factors (R_1 = 0.036, wR_2 = 0.096, S =1.012) Applying this crystallographic information, the elemental composition of the compound could be confirmed (calcd.: C 17.02%, H 2.16%; found: C 16.98%, H 2.19%).[94]

4.6. Modulated Structure[95]

If the three dimensional translation periodicity of a crystal is no longer strictly given, a structure is said to be aperiodic. Among these, incommensurately modulated structures are prominent examples. Hereby atomic positions are not equal in all unit cells of the crystal. Consequently, they can no longer be described by anisotropic displacement parameters. For this reason their positions are modelled with modulation functions. As a mathematical tool, up to three additional dimensions are employed and a superspace description is created. Those allegorise the changes of the atomic coordinates throughout the crystal. In the reciprocal space plot the presence of modulation is indicated by satellite reflections, which occur in addition to the Bragg reflections, which can no longer be indexed by integer numbers in three dimensions. Hence, they are described by fractions in the corresponding directions of the unit cell formed by the main reflections in so-called q-vectors. Not considering these satellite reflections, an average structure can often be gained from a standard least-square refinement. This model usually shows elongated anisotropic displacement parameters or disordered positions. After such satellite reflections had been found for crystals of the coordination polymer $[Cp*_2Mo_2P_2Se_3(CuI)_3(CH_3CN)]_n$ in an experiment carried out in Regensburg, the further processes were performed in collaboration with *M. Dušek, L. Palatinus* and *V. Petříček*.

A view to the reciprocal space plot revealed a non-merohedrally twinned and modulated structure (Figure 23). Several different crystals were processed to find one without a second domain. However, no single crystal could be found, but the exact twin matrix could be determined by CrysAlis software (1 −⅓ 0; 0 −1 0; 0 0 −1) and a crystal with only 0.94(1)% of the second twin was used for refinement ('boden_5_175'). Interestingly, a precise twin indexation showed the unit cell of the second component with a β-angle by two degrees smaller.

After the determination of the q-vector (0 0.588 0.290), the space group was chosen $P\bar{1}(\alpha\beta\gamma)0$ employing Jana2006[96] and the structure was solved by SUPERFLIP. After some least-square refinement cycles, all atomic positions could be approximated, and a view of the average structure revealed an apparently disordered $(Cu_6I_6)(CH_3CN)$ substructure together with an ordered $Cp*_2Mo_2P_2Se_3$ part and a R-value below 4% (Figure 24).

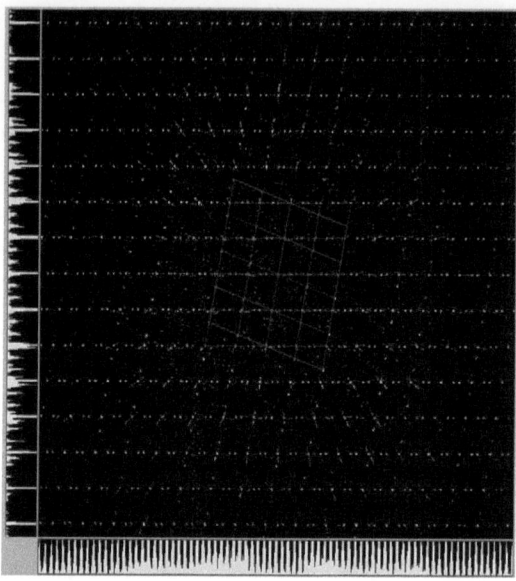

Figure 23: Reciprocal space plot of a crystal of $[Cp*_2Mo_2P_2Se_3(CuI)_3(CH_3CN)]_n$. View along $a*$ of the main component. Lattices of both twin components are shown.

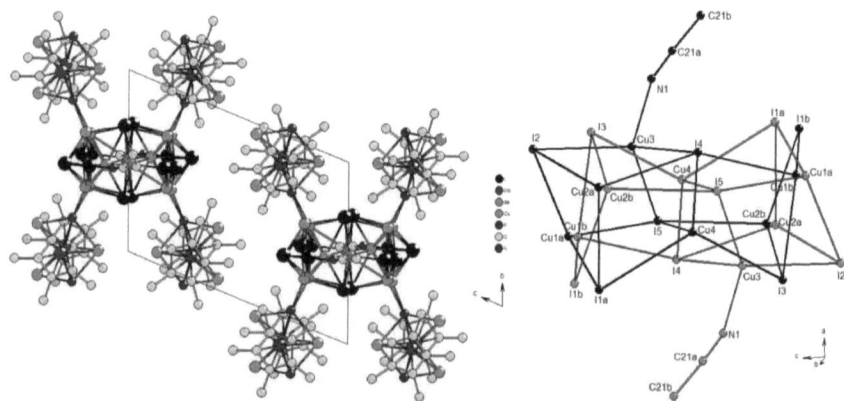

Figure 24: The average structure of 'boden_5_175'. Apparently disordered parts (left). The two cage motifs related by the crystallographic inversion centre in the middle of the copper halide substructure (right).

Consequently, the modelling of the copper iodide cage required further treatment. The assumption that the superimposed positions of the average structure are separated in the four-dimensional description led to the application of discontinuous crenel functions.[97] One of the iodine atoms (I2) was refined with a fixed width of 0.5 for the crenel function which was then used as reference function for the other atoms by defining the interval in the fourth dimension in superspace. This treatment results in two separated cage positions in the modulated model. Both motifs alternately appear and vanish in phase. An additional disorder only affects the two carbon atoms of the acetonitrile ligand. However, the R-values for the modulated structure are unsatisfactory (6.22% for the main reflections and 9.36% for the satellites), since this for the main reflections should improve compared to the average structure. The reason for this observation is found in high residual electron density of about five electrons close to the crenel maxima of the iodine positions. Thus, an additional cage with crenel functions shifted by 0.5 was created. Employing this model, the occupancies refined to 90:10 and the R-values went down to 3.04% for the main reflections and 7.84% for the satellites. An explanation for this behaviour can be found in the second domain visible in the reciprocal space plot. Applying the twin matrix in the refinement does not affect the minor cage position. As mentioned before, the unit cell of this second component exhibits a slightly different unit cell compared to the one indexed by the main reflections. This effect might be caused by the presence of a second phase. In this case, the structure factors would not sum up to result in the observed intensity (twinning), but intensities (multiphase). However, this problem could not be clarified due to an unreliable data reduction.

Finally, a data simulation was carried out to validate the structure model. Since the structure model contains crenel functions, a contribution to the intensities of higher order satellites is possible, but the observed diffraction pattern only shows first order ones. Applying Jana2006, the theoretical intensities of the first and second order satellites were computed. The strongest second order reflections turned out to be twenty times weaker than the first order ones. This information verified the model, since those intensities appeared to be close to the observability border.

5. Experimental Section

5.1. General Methods

All manipulations were performed under an atmosphere of dry argon using standard glovebox and Schlenk techniques. All solvents were freshly distilled from appropriate drying agents immediately prior to use. IR spectra were recorded on a Varian FTS-800 spectrometer. Intensities and positions of the stretches are given by the following abbreviations: w (weak), m (medium), s (strong), vs (very strong) and sh (shoulder). EI-MS spectra were acquired on a Varian MATR 711 mass spectrometer. NMR spectra were gained from a Bruker Avance 400 or on an Avance 300 spectrometer. All signals in the NMR spectra are broad due to the coupling to ^{27}Al nucleus (I = 5/2). For this reason and for reason of very low concentrations, tungsten satellites cannot be reported for all tungsten bound phosphorus signals. Furthermore, the resonances corresponding to the protons at the aluminium atoms are not mentioned for the same reasons. The coupling patterns are indicated by the following abbreviations: s (singlet), d (doublet), t (triplet) and q (quartet). Details for the crystal structure determination procedures are given in chapter 4.1. The starting materials were synthesized according to published methods: [{(CO)$_5$W}PH$_3$],[98] H$_3$Al·NMe$_3$,[99] [(COD)Rh(μ-Cl)]$_2$,[100] P(SiMe$_3$)$_3$,[101] HP(SiMe$_3$)$_2$,[102], H$_2$P(SiMe$_3$)[103], 1,3,4,5-tetramethylimidazolin-2-ylidene[104], ClBH$_2$·NMe$_3$[105], LiSb(SiMe$_3$)$_2$[106]. The syntheses of **1**, **2** and **3** are reported in my diploma thesis.[46] Other chemicals were obtained from Aldrich (LiAlH$_4$) or Merck (AlCl$_3$).

5.2. Alternative Synthesis of [{(CO)$_5$W}H$_2$PAlH$_2$·NMe$_3$] (4)

[{W(CO)$_5$}PH$_3$] (358 mg, 1.00 mmol) and H$_3$Al·NMe$_3$ (89 mg, 1.00 mmol) were dissolved in dichloromethane (10 ml) and stirred at room temperature until gas evolution ceased (1 h). Crystallization at −28 °C gave pale yellow crystals of **4** which were decanted quickly and dried under vacuum. Yield: 364 mg (82%). Analytical data are identical to the initially published ones.[35]

5.3. [{(CO)$_5$W}HPAlH·NMe$_3$]$_3$ (5)

a) Compound **4** (445 mg, 1.00 mmol) was stirred for two hours at 30 °C in toluene (15 ml), which results in a yellow solution from which yellow crystals of **4** were obtained after five days at 4 °C.

b) Compound **4** (445 mg, 1.00 mmol) was dissolved in dichloromethane (10 ml) and stirred for one hour at room temperature. At 4 °C, **5** co-crystallizes with **6** (105 mg, 24%) after six days. The crystals were separated by means of different colour and shape.

Yield: a) 210 mg (47%); b) 100 mg (22%)

^{1}**H NMR** (400.13 MHz, CD$_2$Cl$_2$, 300 K, TMS): δ = 0.26 (d, ^1J(HP) = 242 Hz, 1 H), 0.51 (d, ^1J(HP) = 238 Hz, 1 H), 0.54 (d, ^1J(HP) = 223 Hz, 1 H), 2.84 (s, 18 H, NMe$_3$), 2.86 ppm (s, 9 H, NMe$_3$)

^{1}**H{^{31}P} NMR** (400.13 MHz, CD$_2$Cl$_2$, 300 K, TMS): δ = 0.26 (s, 1 H), 0.51 (s, 1 H), 0.54 (s, 1 H), 2.84 (s, 18 H, NMe$_3$), 2.86 ppm (s, 9 H, NMe$_3$)

^{31}**P NMR** (161.93 MHz, CD$_2$Cl$_2$, 300 K, H$_3$PO$_4$ 85%): δ = –328.5 (d, ^1J(PH) = 242 Hz, 1 P), –328.2 (d, ^1J(PH) = 238 Hz, 1 P), –317.4 ppm (d, ^1J(PH) = 223 Hz, 1 P)

^{31}**P{^{1}H} NMR** (161.93 MHz, CD$_2$Cl$_2$, 300 K, H$_3$PO$_4$ 85%): δ = –328.5 (s, 1 P), –328.2 (s, 1 P), –317.4 ppm (s, 1 P)

IR (KBr): $\tilde{v}$ = 2988 (w), 2940 (w), 2777 (w), 2299 (w, sh, PH), 2276 (w, PH), 2079 (m, CO), 2062 (s, CO), 1915 (vs, CO), 1670 (w, sh, AlH) 1476 (m), 1474 (m), 1412 (w), 1240 (w), 1105 (w), 1013 (m), 989 (w), 816 (w), 714 (w), 689 (w), 650 (w), 600 (m), 561 (w), 451 cm^{-1} (w)

MS (70 eV): m/z (%): 358 (24) [W(CO)$_5$PH$_3$]$^+$, 357 (14) [Al$_3$P$_3$H$_6$(NMe$_3$)$_3$]$^+$, 327 (6) [W(CO)$_4$P]$^+$, 300 (25) [W(CO)$_3$PH]$^+$, 298 (22) [Al$_3$P$_3$H$_6$(NMe$_3$)$_2$]$^+$, 272 (23) [W(CO)$_2$PH]$^+$, 243 (16) [W(CO)$_5$P]$^+$, 215 (9) [WP]$^+$

Crystallographic Data for 5 · CH$_2$Cl$_2$:

Empirical formula	C$_{25}$H$_{35}$Al$_3$Cl$_2$N$_3$O$_{15}$P$_3$W$_3$
Formula weight M	1413.83
Device type	Oxford Diffraction Gemini R Ultra
Crystal colour and shape	yellow parallelepiped
Crystal size	0.12 x 0.11 x 0.07 mm^3
Temperature T	123(1) K
Radiation (λ)	Cu (1.54178 Å)
Crystal system	monoclinic
Space group	$P2_1/n$
Unit cell dimensions	a = 10.1237(1) Å
	b = 22.2498(2) Å β = 91.524(1)°
	c = 20.3792(2) Å
Volume V	4588.80(8) Å^3
Formula units Z	4
Absorption correction type	multi-scan
Absorption coefficient $\mu_{Cu-K\alpha}$	16.739 mm^{-1}
Density (calculated) ρ_{calc}	2.046 g/cm^3
$F(000)$	2664
Theta range θ_{min} / θ_{max}	2.94 / 66.66°
Index ranges	$-11 < h < 11, -25 < k < 21, -24 < l < 23$
Reflections collected	18307
Independent reflections [$I > 2\sigma(I)$]	6766 (R_{int} = 0.0251)
Completeness to full theta	0.958
Transmission T_{min} / T_{max}	0.148 / 0.310
Data / restraints / parameters	7767 / 0 / 514
Goodness-of-fit on F^2 S	1.031
Final R-values [$I > 2\sigma(I)$]	R_1 = 0.0256, wR_2 = 0.0592
Final R-values (all data)	R_1 = 0.0330, wR_2 = 0.0618
Largest difference hole and peak $\Delta\rho$	-0.877, 1.018 eÅ^{-3}
Refinement	Coordinates of H1 to H6 are refined; all others are constrained according to the riding model.

5.4. [⟨{(CO)$_5$W}HPAlH·NMe$_3$⟩$_2$⟨(CO)$_5$WPAl·NMe$_3$⟩] (6)

a) See preparation of **5** as an alternative method for **6** as second product.

b) Compound **5** (200 mg, 0.15 mmol) was crystallized from toluene and dissolved in dichloromethane (20 ml) using an ultrasonic bath (15 min, 35 kHz). The solution was stirred at room temperature for one hour and reduced to a volume of 10 ml. At –28 °C dark yellow crystals of **5** were obtained as well as yellow crystals of **6**.

Yield: 80 mg (40%)

^{1}H NMR (400.13 MHz, CD$_2$Cl$_2$, 300 K, TMS): δ = 0.18 (d, ^1J(HP) = 234 Hz, 1 H), 0.90 (d, ^1J(HP) = 227 Hz, 1 H), 2.80 ppm (s, 27 H, NMe$_3$)

^{1}H{^{31}P} NMR (400.13 MHz, CD$_2$Cl$_2$, 300 K, TMS): δ = 0.18 (s, 1 H), 0.90 (s, 1 H), 2.80 ppm (s, 27 H, NMe$_3$)

^{31}P NMR (161.93 MHz, CD$_2$Cl$_2$, 300 K, H$_3$PO$_4$ 85%): δ = –312.3 (s, 1 P), –289.4 (d, ^1J(HP) = 227 Hz, 1 P), –267.6 ppm (d, ^1J(HP) = 234 Hz, 1 P)

^{31}P{^{1}H} NMR (161.93 MHz, CD$_2$Cl$_2$, 300 K, H$_3$PO$_4$ 85%): δ = –312.3 (s, 1 P), –289.4 (s, 1 P), –267.6 ppm (s, 1 P)

IR (KBr): $\tilde{\nu}$ = 2965 (w), 2933 (w), 2855 (w), 2464 (w), 2274 (w, PH), 2078 (m, CO), 1927 (vs, CO), 1647 (w, AlH), 1262 (w), 1096 (w), 1011 (m), 803 (w), 588 cm^{-1} (w)

MS (70 eV): m/z (%): 358 (67) [W(CO)$_5$PH$_3$]$^+$, 355 (58) [Al$_3$P$_3$H$_4$(NMe$_3$)$_3$]$^+$, 327 (13) [W(CO)$_4$P]$^+$, 300 (69) [W(CO)$_3$PH]$^+$, 272 (67) [W(CO)$_2$PH]$^+$, 243 (45) [W(CO)$_5$P]$^+$, 215 (26) [WP]$^+$

Crystallographic Data for 6:

Empirical formula	$C_{24}H_{31}Al_3N_3O_{15}P_3W_3$
Formula weight M	1326.89
Device type	Oxford Diffraction Gemini R Ultra
Crystal colour and shape	yellow block
Crystal size	0.15 x 0.10 x 0.08 mm^3
Temperature T	123(1) K
Radiation (λ)	Cu (1.54178 Å)
Crystal system	monoclinic
Space group	$P2_1/c$
Unit cell dimensions	a = 20.2444(3) Å
	b = 9.7129(1) Å β = 98.811(1)°
	c = 21.6887(3) Å
Volume V	4214.4(1) Å^3
Formula units Z	4
Absorption correction type	multi-scan
Absorption coefficient $\mu_{Cu-K\alpha}$	17.033 mm^{-1}
Density (calculated) ρ_{calc}	2.091 g/cm^3
$F(000)$	2488
Theta range θ_{min} / θ_{max}	2.21 / 66.63°
Index ranges	$-24 < h < 22, -11 < k < 10, -25 < l < 25$
Reflections collected	17243
Independent reflections [$I > 2\sigma(I)$]	5913 (R_{int} = 0.0342)
Completeness to full theta	0.952
Transmission T_{min} / T_{max}	0.072 / 0.260
Data / restraints / parameters	7105 / 9 / 481
Goodness-of-fit on F^2 S	1.033
Final R-values [$I > 2\sigma(I)$]	R_1 = 0.0396, wR_2 = 0.0982
Final R-values (all data)	R_1 = 0.0494, wR_2 = 0.1028
Largest difference hole and peak $\Delta\rho$	-2.805, 3.700 eÅ^{-3} (located close to tungsten atoms)
Refinement	H1 to H4 positionally refined with same P–H, Al–H distances (SADI). C8 and O8 restrained to similar displacement parameters (SIMU, DELU)

5.5. Alternative Synthesis of [{(CO)$_5$WPH$_2$}(Me$_3$N)AlPH{W(CO)$_5$}]$_2$ (7)

As *Vogel* reported, **6** can be synthesized in moderate yields either with or without a catalyst.[45] Alternative reaction conditions also resulted in **6**. a) [{(CO)$_5$W}PH$_3$] (179 mg, 0.50 mmol), H$_3$Al·NMe$_3$ (45 mg, 0.50 mmol) and [(COD)Rh(μ-Cl)]$_2$ (12 mg, 5 mol-%) were dissolved in CH$_2$Cl$_2$ and stirred until gas evolution ceased (1 h). From the orange solution, pale yellow crystals were obtained as well as crystals of **4** and **5**. b) To a dichloromethane solution (2 ml) of [{(CO)$_5$W}PH$_3$] (358 mg, 1.00 mmol) a less concentrated solution of H$_3$Al·NMe$_3$ (45 mg, 0.50 mmol) in dichloromethane (10 ml) was slowly added and refluxed until gas evolution ceased (1 h). After storing the solution at 4 °C for three days, pale yellow crystals of **6** were obtained. In addition to the analytical data reported by *Vogel*, NMR chemical shifts could be obtained from the reaction mixture.

Yield: a) 88 mg (11%), b) 110 mg (27%)

^{1}**H NMR** (400.13 MHz, CD$_2$Cl$_2$, 300 K, TMS): δ = 0.66 (d, ^1J(HP) = 229 Hz, 2 H), 2.82 (s, 18 H, NMe$_3$), 2.83 ppm (d, ^1J(HP) = 287 Hz, 4 H)

^{31}**P NMR** (161.93 MHz, CD$_2$Cl$_2$, 300 K, H$_3$PO$_4$ 85%): δ = –287.6 (d, ^1J(HP) = 229 Hz, 2 P), –234.6 ppm (t, ^1J(HP) = 287 Hz, 2 P)

^{31}**P{^{1}H} NMR** (161.93 MHz, CD$_2$Cl$_2$, 300 K, H$_3$PO$_4$ 85%): δ = –287.6 (s, 2 P), –234.6 ppm (s, 2 P)

5.6. [{(CO)$_5$WPH$_2$}(Me$_2$EtN)AlPH{W(CO)$_5$}]$_2$ (10) and [⟨{W(CO)$_5$}HPAl(Me$_2$EtN)⟩$_2$ μ-⟨{(CO)$_5$WPH}$_2$Al(Me$_2$EtN)⟩] (11)

To a solution of 179 mg (0.5 mmol) [{(CO)$_5$W}PH$_3$], 1.1 ml of a 0.5 mol/l solution of H$_3$Al·NMe$_2$Et in toluene was added. Following gas evolution, the reaction mixture was stirred for an additional 30 minutes. The yellow-orange solution is stored at –28 °C and pale yellow crystals of **10** and **11** were obtained after six months. Only a few crystals could be isolated. Analytical data cannot be provided, due to the lack of substance and the insolubility of the compounds in common non-coordinating organic solvents. Furthermore, the crystals could only be separated by determining the unit cell in X-ray experiments.

Crystallographic Data for 10:

Empirical formula	$C_{28}H_{28}Al_2N_2O_{20}P_4W_4$
Formula weight M	1625.72
Device type	Oxford Diffraction Gemini R Ultra
Crystal colour and shape	pale yellow block
Crystal size	0.17 x 0.15 x 0.14 mm^3
Temperature T	123(1) K
Radiation (λ)	Cu (1.54178 Å)
Crystal system	triclinic
Space group	$P\bar{1}$
Unit cell dimensions	a = 10.3963(5) Å, α = 64.540(4)°
	b = 11.1880(5) Å, β = 88.338(4)°
	c = 11.8955(5) Å, γ = 72.518(4)°
Volume V	1183.5(1) Å^3
Formula units Z	1
Absorption correction type	multi-scan
Absorption coefficient $\mu_{Cu-K\alpha}$	19.783 mm^{-1}
Density (calculated) ρ_{calc}	2.281 g/cm^3
$F(000)$	752
Theta range θ_{min} / θ_{max}	4.14 / 62.34°
Index ranges	$-11 < h < 11, -12 < k < 12, -13 < l < 13$
Reflections collected	16785
Independent reflections $[I > 2\sigma(I)]$	2986 (R_{int} = 0.0419)
Completeness to full theta	0.985
Transmission T_{min} / T_{max}	0.528 / 1.000
Data / restraints / parameters	3698 / 3 / 286
Goodness-of-fit on F^2 S	1.074
Final R-values $[I > 2\sigma(I)]$	R_1 = 0.0448, wR_2 = 0.1117
Final R-values (all data)	R_1 = 0.0580, wR_2 = 0.1214
Largest difference hole and peak $\Delta\rho$	−0.575, 2.902 eÅ^{-3} (located close to tungsten atoms)
Refinement	H1, H2A, H2B refined with fixed distances (DFIX).

Crystallographic Data for 11 · C_7H_8:

Empirical formula	$C_{39}H_{46}Al_3N_3O_{20}P_4W_4$
Formula weight M	1816.97
Device type	Oxford Diffraction Gemini R Ultra
Crystal colour and shape	pale yellow block
Crystal size	0.18 x 0.18 x 0.12 mm^3
Temperature T	123(1) K
Radiation (λ)	Cu (1.54178 Å)
Crystal system	triclinic
Space group	$P\bar{1}$
Unit cell dimensions	a = 11.7356(4) Å, α = 99.999(2)°
	b = 12.2754(4) Å, β = 101.488(2)°
	c = 22.0653(6) Å, γ = 103.550(2)°
Volume V	2946.3(2) Å^3
Formula units Z	2
Absorption correction type	multi-scan
Absorption coefficient $\mu_{Cu-K\alpha}$	16.123 mm^{-1}
Density (calculated) ρ_{calc}	2.048 g/cm^3
$F(000)$	1712
Theta range θ_{min} / θ_{max}	3.90 / 62.16°
Index ranges	$-13 < h < 13, -14 < k < 14, -25 < l < 24$
Reflections collected	21354
Independent reflections [$I > 2\sigma(I)$]	8196 (R_{int} = 0.0404)
Completeness to full theta	0.972
Transmission T_{min} / T_{max}	0.267 / 1.000
Data / restraints / parameters	9046 / 60 / 720
Goodness-of-fit on F^2 S	1.063
Final R-values [$I > 2\sigma(I)$]	R_1 = 0.0489, wR_2 = 0.1254
Final R-values (all data)	R_1 = 0.0540, wR_2 = 0.1298
Largest difference hole and peak $\Delta\rho$	−3.362, 3.240 eÅ^{-3} (located close to tungsten atoms)
Refinement	Disorder modelled applying constraints (EADP, EXYZ) and restraints (ISOR); Al–H and P–H distances fixed (DFIX).

5.7. Synthesis of (Me$_3$Si)$_2$PAlH$_2$·NMe$_3$ (12)

To a solution of 89 mg (1 mmol) H$_3$Al·NMe$_3$ in 5 ml dichloromethane or toluene, 0.22 ml (1 mmol) HP(SiMe$_3$)$_2$ were added and the reaction mixture was stirred for two hours at room temperature. After the gas evolution ceased the solvent is reduced until the solution became cloudy. The solution became clear after re-warming to room temperature. The product crystallized at 4 °C within a few hours in colourless blocks. The crystals were extremely sensitive at ambient temperatures and had to be stored at −28 °C to avoid decomposition. For that reason MS and IR spectroscopy could not be carried out.

Yield: ca. 50% in both solvents (determined by NMR)

^{1}H NMR (400.13 MHz, CD$_2$Cl$_2$, 300 K, TMS): δ = 0.55 (dm, ^1J(HP) = 4.4 Hz, 36 H, SiMe$_3$), 1.86 (s, 9 H, NMe$_3$), 4.5 ppm (s, 2 H, AlH$_2$)

^{31}P NMR (161.93 MHz, CD$_2$Cl$_2$, 300 K, H$_3$PO$_4$ 85%): δ = −283.3 ppm (s, P(SiMe$_3$)$_2$)

^{31}P{^{1}H} NMR (161.93 MHz, CD$_2$Cl$_2$, 300 K, H$_3$PO$_4$ 85%): δ = −283.3 ppm (s, P(SiMe$_3$)$_2$)

If the solvent was removed under reduced pressure from the toluene reaction mixture, colourless crystals of [(Me$_3$Si)$_2$PAlH$_2$]$_3$ could be isolated. The unit cell parameters and the NMR shifts are identical to those reported by *Wells* and *White*.[64]

5. EXPERIMENTAL SECTION

Crystallographic Data for 12:

Empirical formula	$C_9H_{29}AlNPSi_2$
Formula weight M	265.46
Device type	Oxford Diffraction Gemini R Ultra
Crystal colour and shape	colourless block
Crystal size	0.24 x 0.17 x 0.15 mm^3
Temperature T	120(1) K
Radiation (λ)	Cu (1.54178 Å)
Crystal system	orthorhombic
Space group	$Cmc2_1$
Unit cell dimensions	a = 14.2568(4) Å
	b = 11.0532(3) Å
	c = 11.1594(3) Å
Volume V	1758.53(8) Å^3
Formula units Z	4
Absorption correction type	multi-scan
Absorption coefficient $\mu_{Cu-K\alpha}$	2.966 mm^{-1}
Density (calculated) ρ_{calc}	1.003 g/cm^3
$F(000)$	584
Theta range θ_{min} / θ_{max}	5.06 / 51.57°
Index ranges	$-14 < h < 11, -11 < k < 11, -11 < l < 11$
Reflections collected	2985
Independent reflections [$I > 2\sigma(I)$]	867 (R_{int} = 0.0229)
Completeness to full theta	0.996
Transmission T_{min} / T_{max}	0.416 / 1.000
Data / restraints / parameters	941 / 1 / 78
Flack parameter x	0.11(4)
Goodness-of-fit on F^2 S	1.051
Final R-values [$I > 2\sigma(I)$]	R_1 = 0.0250, wR_2 = 0.0594
Final R-values (all data)	R_1 = 0.0278, wR_2 = 0.0606
Largest difference hole and peak $\Delta\rho$	$-0.138, 0.179$ eÅ^{-3}
Refinement	H1 refined without restraints.

5.8. Synthesis of [(Me$_3$Si)PAlH·NMe$_3$]$_2$ (13)

A solution of 644 mg (7.23 mmol) H$_3$Al·NMe$_3$ in 5 ml CH$_2$Cl$_2$ and 1 ml (7.23 mmol) H$_2$P(SiMe$_3$) was slowly added while stirring. Already during the solvent reducing procedure colourless blocks formed, which decomposed quickly at room temperature turning white from the surface. However, from a reaction carried out in a NMR tube spectroscopic data could be gained.

Yield: ca. 65% (determined by NMR)

^{1}**H NMR** (400.13 MHz, CD$_2$Cl$_2$, 300 K, TMS): δ = 0.05 (s, 18 H, SiMe$_3$), 3.32 (s, 18 H, NMe$_3$)

^{31}**P NMR** (161.93 MHz, CD$_2$Cl$_2$, 300 K, H$_3$PO$_4$ 85%): δ = –284.5 ppm (s, P(SiMe$_3$))

^{31}**P{^{1}H} NMR** (161.93 MHz, CD$_2$Cl$_2$, 300 K, H$_3$PO$_4$ 85%): δ = –284.5 ppm (s, P(SiMe$_3$))

5.9. Synthesis of H$_3$Al·NHC^{Me} (14)

The 1,3,4,5-tetramethylimidazolin-2-ylidene was purified by sublimation beforehand the synthesis. 460 mg (12.1 mmol) LiAlH$_4$ were suspended in 20 ml diethylether and cooled to –50 °C. A solution of 540 mg (4.05 mmol) AlCl$_3$ in 10 ml diethylether at –50 °C was added. The solution was allowed to warm up to –40 °C and beforehand the addition of 2.01 g (16.2 mmol) carbene. The reaction mixture was stirred for 75 minutes. Thereby, the solution was warmed to –20 °C. After the filtration from LiCl at 0 °C the colourless solution was stored at –28 °C to obtain colourless needles of **14**. The pure product decomposes at ambient temperatures and has to be stored at –28 °C.

Yield: 1.25 g (50%)

^{1}**H NMR** (400.13 MHz, CD$_2$Cl$_2$, 300 K, TMS): δ = 1.14 (s, 6 H, N–Me), 3.19 (s, 9 H, C–Me), 4.55 (s, 3 H, AlH$_3$)

Crystallographic Data for 14:

Empirical formula	$C_7H_{15}AlN_2$
Formula weight M	154.19
Device type	Oxford Diffraction Gemini R Ultra
Crystal colour and shape	colourless rod
Crystal size	0.50 x 0.07 x 0.04 mm^3
Temperature T	123(1) K
Radiation (λ)	Cu (1.54178 Å)
Crystal system	monoclinic
Space group	$P2_1/c$
Unit cell dimensions	a = 7.5394(3) Å
	b = 15.5295(5) Å β = 103.038(4)°
	c = 8.4999(3) Å
Volume V	969.54(6) Å^3
Formula units Z	4
Absorption correction type	multi-scan
Absorption coefficient $\mu_{Cu\text{-}K\alpha}$	1.322 mm^{-1}
Density (calculated) ρ_{calc}	1.056 g/cm^3
$F(000)$	336
Theta range θ_{min} / θ_{max}	5.70 / 66.69°
Index ranges	$-8 < h < 7, -18 < k < 18, -10 < l < 10$
Reflections collected	8487
Independent reflections [$I > 2\sigma(I)$]	1570 (R_{int} = 0.0215)
Completeness to full theta	0.987
Transmission T_{min} / T_{max}	0.705 / 1.000
Data / restraints / parameters	1687 / 0 / 96
Goodness-of-fit on F^2 S	1.089
Final R-values [$I > 2\sigma(I)$]	R_1 = 0.0533, wR_2 = 0.1426
Final R-values (all data)	R_1 = 0.0558, wR_2 = 0.1437
Largest difference hole and peak $\Delta\rho$	$-0.272, 0.370$ eÅ^{-3}
Refinement	Coordinates of H1 and H2 refined with equal bond lengths (AFIX 134).

5.10. Synthesis of (Me$_3$Si)$_2$SbBH$_2$·NMe$_3$ (15)

369 mg (1 mmol) LiSb(SiMe$_3$)$_2$·DME and 113 mg (1 mmol) ClBH$_2$·NMe$_3$ were dissolved in 20 ml *n*-hexane. The mixture was refluxed at 75 °C for 18 hours. The black precipitate was removed by filtration over diatomaceous earth resulting in a yellow solution. The ^{1}H NMR spectrum of the crude reaction mixture is inconclusive due to overlapping signals.

^{11}B NMR (128.38 MHz, CD$_2$Cl$_2$, 300 K, TMS): δ = –8.5 ppm (t, ^1J(BH) = 114.8 Hz, BH$_2$)

^{11}B{^{1}H} NMR (128.38 MHz, CD$_2$Cl$_2$, 300 K, TMS): δ = –8.5 ppm (s, BH$_2$)

6. Summary and Conclusions

6.1. Synthetic Results

During my diploma thesis the compounds [{(CO)$_5$W}RPAlH·NR'$_3$]$_2$ (**2**: R = H, R' = Et; **3**: R = Ph, R' = Me) were synthesized, and the first evidence for isomerization processes in solution was found in the NMR spectra.[46] Within this work the isomerization mechanisms could be clarified by applying theoretical computations, which showed an amine base exchange via a S$_N$2 reaction pathway as most probable explanation (Scheme 12). The theoretical considerations show an easy cleavage of the dative N–Al bond upon geometry optimization.

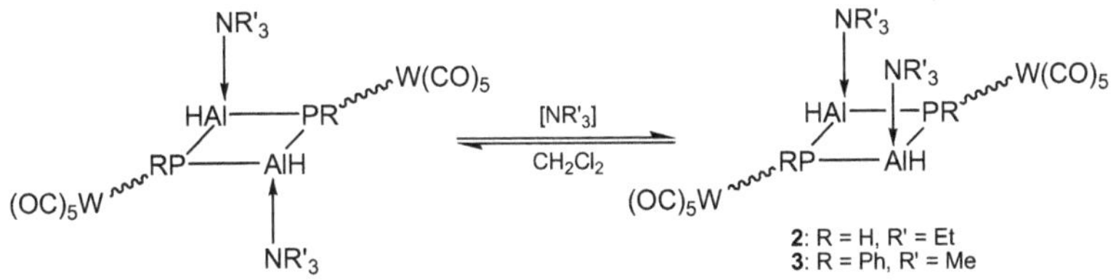

Scheme 12: Isomerization equilibrium of **2** and **3** in dichloromethane solution.

For the trimethylamine derivative [{(CO)$_5$W}HPAlH·NMe$_3$] the synthesis of the starting material **4** was optimized to result in an almost doubled yield compared to the previously reported method.[34] This increase was achieved by changing the reaction conditions, in particular solvent and temperature. This observation was further extended to control the subsequent oligomerization of **4**. The use of toluene as the solvent lead to the exclusive formation of the trimerization product [{(CO)$_5$W}HPAlH·NMe$_3$]$_3$ (**5**). In contrast, the reaction in dichloromethane also resulted in **5** but together with [{(CO)$_5$W}HPAlH·NMe$_3$]$_2$[(CO)$_5$WPAl·NMe$_3$] (**6**). It could be shown that the hydrogen elimination leading to the formation of **6** can be supported by treating a solution of **5** with ultrasound. As a further side-product [{(CO)$_5$WPH$_2$}(Me$_3$N)AlPH{W(CO)$_5$}]$_2$ (**7**) was obtained from the reaction of [{(CO)$_5$W}PH$_3$] with H$_3$Al·NMe$_3$ in dichloromethane. If this reaction is followed by NMR spectroscopy, the presence of a four-membered ring intermediate [{(CO)$_5$W}HPAlH·NMe$_3$]$_2$ (**8**) could be proven. It showes similar isomerization behaviour in solution to **2** and **3**, but could not be isolated for reasons of further reactivity towards **6** by addition of another equivalent of **4**. The first assumption that the formation of **7** is a result of a reaction of the four-membered intermediate with two equivalents of

6. SUMMARY AND CONCLUSIONS

[{(CO)$_5$W}PH$_3$], could be ruled out as energetically unfavourable. However, it could be shown that another intermediate [{(CO)$_5$W}PH$_2$]$_2$AlH·NMe$_3$ (**9**), also found in the NMR spectrum of the crude reaction mixture, gives **7** via subsequent hydrogen elimination. Employing the correct stoichiometry together with slow addition of the alane to a concentrated solution of the phosphane complex allowed the yield of **7** to be maximized. Compounds **5** and **6** were fully characterized, and for **7** additional NMR data were gained. An overview of the controlled oligomerization including all intermediates is depicted in Scheme 13.

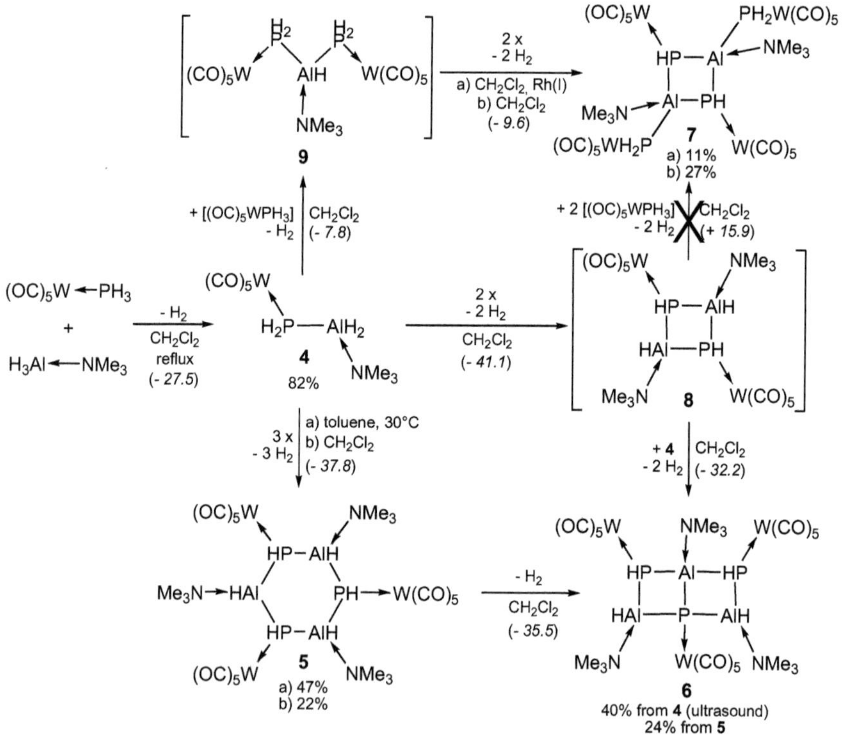

Scheme 13: Possible reaction pathways depending on the reaction conditions. If not further specified all reactions were carried out at room temperature (*Gibbs energies in kJ mol^{-1}*).

By changing the Lewis base to the asymmetric ethyldimethylamine, a previously unknown[63] structural motif for [{W(CO)$_5$}HPAl(NMe$_2$Et)]$_2$μ-[{(CO)$_5$WPH}$_2$Al(NMe$_2$Et)] (**11**) was found in addition to the NMe$_2$Et analogue of **7** [{(CO)$_5$WPH$_2$}(Me$_2$EtN)AlPH{W(CO)$_5$}]$_2$ (**10**) by X-ray structure analysis. Interestingly, no evidence for the monomeric derivative of

6. SUMMARY AND CONCLUSIONS

[{(CO)$_5$W}H$_2$PAlH$_2$·NR$_3$] (R = Me, Et) was found as an intermediate. A self-assembly from [{(CO)$_5$W}PH$_3$] and H$_3$Al·NMe$_2$Et under very slow hydrogen elimination leads to the formation of **10** and **11** instead (Scheme 14).

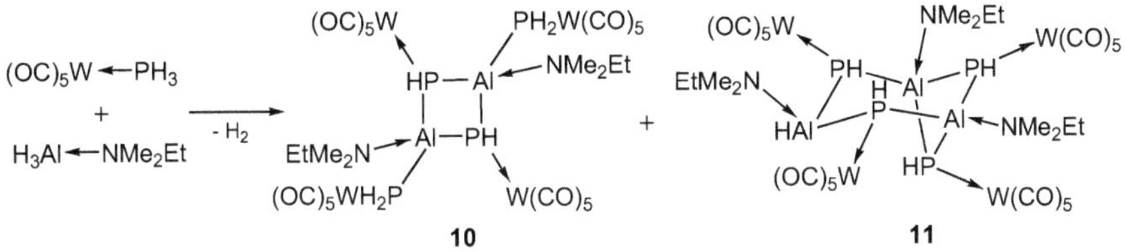

Scheme 14: Synthesis and structural motifs of **10** and **11**.

More interesting reactivity of phosphanylalanes could be shown employing Lewis-acid-free trimethylsilyl substituted phosphanes (Me$_3$Si)PRH (R = Me$_3$Si, H). The reaction of (Me$_3$Si)$_2$PH and H$_3$Al·NMe$_3$ results in the thermally unstable compound (Me$_3$Si)$_2$PAlH$_2$·NMe$_3$ (**12**), both in toluene and CH$_2$Cl$_2$. Compound **12** was characterized by NMR and X-ray structure analysis. A subsequent trimerization towards [(Me$_3$Si)$_2$PAlH$_2$]$_3$ was proven by its unit cell parameters and NMR chemical shifts.[64] The amine base elimination can be considered facile, occuring already at reduced pressures. The synthesis of [(Me$_3$Si)PAlH·NMe$_3$]$_2$ (**13**) from (Me$_3$Si)PH$_2$ and H$_3$Al·NMe$_3$ was also found. Compound **13** appeared to be even less stable than **12**, but could be characterized by NMR spectroscopy.

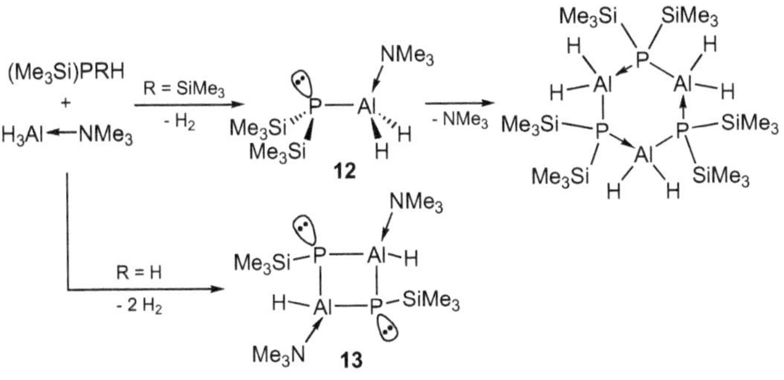

Scheme 15: Reaction pathways leading to the formations of **12**, **13** and the NMe$_3$ elimination product [(Me$_3$Si)$_2$PAlH$_2$]$_3$.

By reacting *in situ* generated H$_3$Al·OEt$_2$ with 1,3,4,5-tetramethylimidazolin-2-ylidene (NHC^{Me})[65] a new Lewis base could be introduced (Scheme 16). By this way formed H$_3$Al·NHC^{Me} was characterized by X-ray structure analysis and ^{1}H NMR spectroscopy and can be synthesized in large scales and good yields. Hence, it might be a good synthon for future investigations.

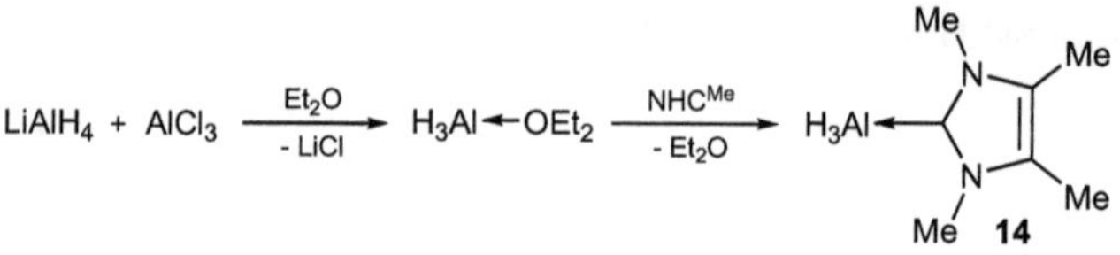

$$LiAlH_4 + AlCl_3 \xrightarrow[-LiCl]{Et_2O} H_3Al \leftarrow OEt_2 \xrightarrow[-Et_2O]{NHC^{Me}} H_3Al \leftarrow \underset{\underset{50\%}{\mathbf{14}}}{NHC^{Me}}$$

Scheme 16: Synthesis of **14** from LiAlH$_4$, AlCl$_3$ and NHC^{Me}.

Extending the research scope of the pentelylboranes to the heavier group 15 homologues antimony and bismuth led to the synthesis of (Me$_3$Si)$_2$SbBH$_2$·NMe$_3$ (**15**) as first step towards the formation of H$_2$SbBH$_2$·NMe$_3$. However, the subsequent methanolysis causes decomposition (Scheme 17). In the case of its bismuth derivative already the salt elimination is unsuccessful and results in elemental bismuth, among other decomposition products. Employing theoretical investigations the observed instability could be confirmed by strongly exergonic decomposition reactions.

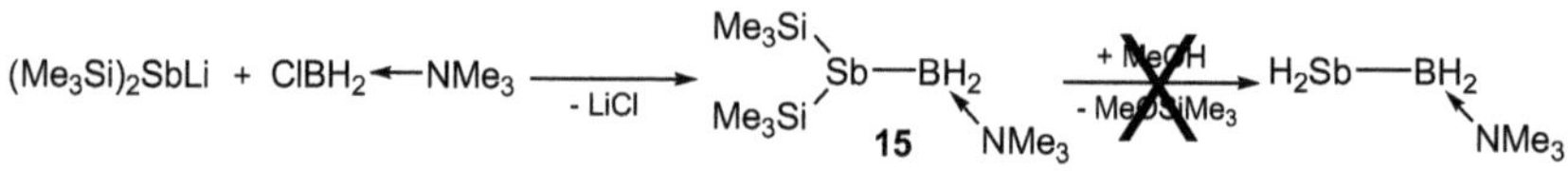

$$(Me_3Si)_2SbLi + ClBH_2 \leftarrow NMe_3 \xrightarrow{-LiCl} \underset{\mathbf{15}}{\underset{Me_3Si}{\overset{Me_3Si}{Sb-BH_2 \leftarrow NMe_3}}} \xrightarrow[-MeOSiMe_3]{+MeOH} \cancel{} \; H_2Sb-BH_2 \leftarrow NMe_3$$

Scheme 17: Synthesis of **15**.

6.2. Crystallographic Results

During the work to this thesis about 200 crystal samples were processed. Not taking starting materials, side-products, known compounds and structures finalized by *Drs Zabel, Virovets* and *Peresypkina* at the beginning of this work into account, 132 crystal structures could be determined. Among those 51 turned out to be disordered (35 main part, 4 solvent/anion/cation, 12 both) in 9 cases SQUEEZE was applied to severely disordered solvent molecules. 16 structures were twinned (8 merohedrally, 3 pseudo-merohedrally, 5 non-merohedrally). One incommensurately modulated structure could be solved in cooperation with *Drs Dušek, Palatinus* and *Prof. Dr. Petříček*. Furthermore, synchrotron radiation experiments were carried out at the ANKA beamline in Karlsruhe together with *Dr Balázs*. In addition to this, technical services were executed (tube changes; (re-)alignments of the X-ray beam; detector and Dewar vessel evacuations; a computer exchange and re-setup; diverse filter exchanges, leak repairings, computer-, software- and connectivity problems). Furthermore, the necessary preparations and the new installing of the SuperNova device were

6. SUMMARY AND CONCLUSIONS 77

supervised in absence of *Dr Zabel*. In cooperation with *Dr Meyer* from Agilent Technologies (formerly known as Oxford Diffraction Ltd.) new features could be implemented into the CrysAlis software and several software problems could be solved. During workshops and visits in Karlsruhe (synchrotron experiments at the ANKA, 2008), Oxford ("Oxford Diffraction User Meeting", 2008 and 2009), Freiburg ("ChemKrist Workshop 2009"), Prague (collaboration and advanced training on modulated structures, 2009 and 2010) Pécs (COST meeting "Models, Structures, Spectroscopies – Contemporary Methods", 2010), Wroclaw (collaboration on software issues and visit to the Agilent production facilities, 2010) Nancy ("Summer Schools on Mathematical and Topological Crystallography", 2010) and Darmstadt ("26th European Crystallographic Meeting" and "Agilent User Meeting", 2010) the theoretical and technical knowledge could be improved.

7. Appendix

7.1. Supporting Data

The crystallographic data for the 132 determined structures can be obtained upon request (michael.bodensteiner@ur.de). The following files are given for each dataset:

1.hkl	reflection data file processed by XPREP or GRAL
1.res	final result file of the least-square refinement
name.hkl	raw reflection file of the data reduction
name.p4p	unit cell parameter and radiation wavelength input file for XPREP
name.cif	data collection parameter file of the data reduction
*name*_fin.cif	final crystallographic information file
checkcif_*name*.pdf	checkCIF/PLATON report of a basic structural check[107]
squeeze.hkl	optional reflection file if SQUEEZE was applied

Furthermore, details of the theoretical computations of the chapters 3.1 and 3.2 are also available upon email request.

7.2. List of Compounds

1 $[\{(CO)_5W\}H_2PAlH_2 \cdot NEt_3]$
2 $[\{(CO)_5W\}HPAlH \cdot NEt_3]_2$
3 $[\{(CO)_5W\}PhPAlH \cdot NMe_3]_2$
4 $[\{(CO)_5W\}H_2PAlH_2 \cdot NMe_3]$
5 $[\{(CO)_5W\}HPAlH \cdot NMe_3]_3$
6 $[\langle\{(CO)_5W\}HPAlH \cdot NMe_3\rangle_2\langle(CO)_5WPAl \cdot NMe_3\rangle]$
7 $[\{(CO)_5WPH_2\}(Me_3N)AlPH\{W(CO)_5\}]_2$
8 $[\{(CO)_5W\}HPAlH \cdot NMe_3]_2$
9 $[\{(CO)_5W\}PH_2]_2AlH \cdot NMe_3$
10 $[\{(CO)_5WPH_2\}(Me_2EtN)AlPH\{W(CO)_5\}]_2$
11 $[\langle\{W(CO)_5\}HPAl(Me_2EtN)\rangle_2\mu\text{-}\langle\{(CO)_5WPH\}_2Al(Me_2EtN)\rangle]$
12 $(Me_3Si)_2PAlH_2 \cdot NMe_3$
13 $[(Me_3Si)PAlH \cdot NMe_3]_2$
14 $H_3Al \cdot NHC^{Me}$
15 $(Me_3Si)_2SbBH_2 \cdot NMe_3$

7.3. List of Abbreviations

3D	3-dimensional	
Å	Ångström (10^{-10} m)	
cat	catalyst	
CCD	charge-coupled device	
cm^{-1}	reciprocal centimetres	
COD	1,5-*cyclo*-octadiene	C_8H_{12}
Cp	*cyclo*-pentadienyl	η^5-$C_5H_5^-$
Cp*	*penta*-methyl-*cyclo*-pentadiene *or* -pentadienyl	$-C_5(CH_3)_5$, η^5-$C_5(CH_3)_5^-$
CSD	Cambridge Structural Database	
d	spacing between lattice planes	
DFT	density functional theory	
dmap	*N,N*-dimethylaminopyridine	$NC_5H_4N(CH_3)_2$
E	pentel	N, P, As, Sb, Bi
E	energy or normalized structure factors	
E'	triel	B, Al, Ga, In, Tl
Et	ethyl	$-CH_2CH_3$
exp	exponent	
FLP	frustrated Lewis pair	
FLP	frustrated Lewis pair	
G	Gibbs energy	
G^6	6-dimensional space	
H	enthalpy	
h	hour(s)	
hkl	Miller indices	
Hz	Hertz (s^{-1})	
I	intensity ($\sim F^2$)	
i	imaginary part of a complex number	
^{i}Bu	*iso*-butyl	$CH_2CH(CH_3)_2$
^{i}Pr	*iso*-propyl	$-CH(CH_3)_2$
IR	infrared	
J	coupling constant	
K	Kelvin	
kJ	Kilojoule	

λ	wavelength	
LA	Lewis acid	
LB	Lewis base	
LED	light emitting diode	
lm	Lumen (radiant flux)	
μ	dipole moment *or* bridging atom or group	
M	metal	
m/z	mass-to-charge ratio	
Me	methyl	$-CH_3$
Mes	mesityl	$2,4,6-Me_3C_6H_2$
Mes*	2,4,6-*tert*-butylphenyl	$-C_6H_2(C(CH_3)_3)_3$
mol	mole (amount of substance)	
MS	mass spectrometry	
n	integer	
$\tilde{v}$	wave number	
N	number of reflections	
NHC^{Me}	1,3,4,5-tetramethylimidazolin-2-ylidene	$CN_2C_2(CH_3)_4$
NMR	nuclear magnetic resonance	
π	circle number (3.14159265)	
Ph	phenyl	$-C_6H_5$
ppm	parts per million	
Py	pyridine	C_5H_5N
S	entropy	
Σ	sum	
$\sigma(I)$	standard deviation of the intensity	
sin	sine	
S_N	nucleophilic substitution	
^tBu	*tert*-butyl	$-C(CH_3)_3$
θ	dihedral angle *or* diffraction angle	
THF	tetrahydrofurane	C_4H_8O
TMS	tetrametylsilane	$Si(CH_3)_4$
V	volume	
W	Watt	
X	halogen	Cl, Br, I

8. Literature

[1] J. E. Huheey, E. A. Keiter, R. L. Keiter, *Anorganische Chemie: Prinzipien von Struktur und Reaktivität*, 2nd ed., Walter de Gruyter Verlag, Berlin, **1995**.
[2] M. Corso, W. Auwärter, M. Muntwiler, A. Tamai, T. Greber, J. Osterwalder, *Science* **2004** *303*, 217–220.
[3] N. G. Chopra, R. J. Luyken, K. Cherrey, V. H. Crespi, M. L. Cohen, S. G. Louie, A. Zettl, *Science* **1995**, *269*, 966–967.
[4] F. A. Holleman, E. Wiberg, N. Wiberg, *Lehrbuch der Anorganischen Chemie*, 102nd ed., Walter de Gruyter, Berlin, **2007**.
[5] Z. I. Alferov, *Nobel Lecture*, **2000**.
[6] H. Kroemer, *Nobel Lecture*, **2000**.
[7] W. Shockley, H. J. Queisser, *J. Appl. Phys.* **1961**, *32*, 510–519.
[8] D. Guimard, R. Morihara, D. Bordel, K. Tanabe, Y. Wakayama, M. Nishioka, Y. Arakawa, *Appl. Phys. Lett.* **2010**, *96*, 203507.
[9] S. Bush, www.electronicsweekly.com/Articles/2008/11/20/41947/LED-technology-White-LEDs.htm.
[10] T. B. Marder, *Angew. Chem. Int. Ed.* **2007**, *46*, 8116–8118.
[11] A. Staubitz, A. P. M. Robertson, I. Manners, *Chem. Rev.* **2010**, *110*, 4079–4124.
[12] F. H. Stephens, R. T. Baker, M. H. Matus, D. J. Grant, D. A. Dixon, *Angew. Chem.-Int. Edit.* **2007**, *46*, 746–749.
[13] S.-K. Kim, W.-S. Han, T.-J. Kim, T.-Y. Kim, S. W. Nam, M. Mitoraj, L. Pieko, A. Michalak, S.-J. Hwang, S. O. Kang, *J. Am. Chem. Soc.* **2010**, *132*, 9954–9955.
[14] B. L. Conley, T. J. Williams, *Chem. Commun.* **2010**, *46*, 4815–4817.
[15] P. M. Zimmerman, A. Paul, C. B. Musgrave, *Inorg. Chem.* **2009**, *48*, 5418–5433.
[16] A. Staubitz, A. P. Soto, I. Manners, *Angew. Chem. Int. Ed.* **2008**, *47*, 6212–6215.
[17] Ö. Metin, V. Mazumder, S. Özkar, S. Sun, *J. Am. Chem. Soc.* **2010**, *132*, 1468–1469.
[18] J.-M. Yan, X.-B. Zhang, S. Han, H. Shioyama, Q. Xu, *Angew. Chem. Int. Ed.* **2008**, *47*, 2287–2289.
[19] D. W. Himmelberger, L. R. Alden, M. E. Bluhm, L. G. Sneddon, *Inorg. Chem.* **2009**, *48*, 9883–9889.
[20] S. J. Geier, T. M. Gilbert, D. W. Stephan, *J. Am. Chem. Soc.* **2008**, *130*, 12632–12633.
[21] G. C. Welch, R. R. S. Juan, J. D. Masuda, D. W. Stephan, *Science* **2006**, *314*, 1124–1126.
[22] D. W. Stephan, G. Erker, *Angew. Chem. Int. Ed.* **2010**, *49*, 46–76.
[23] H. Dorn, R. A. Singh, J. A. Massey, A. J. Lough, I. Manners, *Angew. Chem. Int. Ed.* **1999**, *38*, 3321–3323.
[24] H. Dorn, R. A. Singh, J. A. Massey, J. M. Nelson, C. A. Jaska, A. J. Lough, I. Manners, *J. Am. Chem. Soc.* **2000**, *122*, 6669–6678.
[25] A. Adolf, *Dissertation*, Regensburg, **2007**.
[26] H. Westenberg, J. C. Slootweg, A. Hepp, J. Kösters, S. Roters, A. W. Ehlers, K. Lammertsma, W. Uhl, *Organometallics* **2010**, *29*, 1323–1330.
[27] T. Habereder, H. Nöth, R. T. Paine, *Eur. J. Inorg. Chem.* **2007**, 4298–4305.
[28] R. J. Wehmschulte, K. Ruhlandt-Senge, P. P. Power, *Inorg. Chem.* **1994**, *33*, 3205–3207.
[29] T. L. Allen, W. H. Fink, *Inorg. Chem.* **1992**, *31*, 1703–1705.
[30] T. L. Allen, A. C. Scheiner, H. F. Schaefer, *Inorg. Chem.* **1990**, *29*, 1930–1936.
[31] M. B. Coolidge, W. T. Borden, *J. Am. Chem. Soc.* **1990**, *112*, 1704–1706.
[32] H.-J. Himmel, *Dalton Trans.* **2003**, 3639–3649.
[33] H.-J. Himmel, *Eur. J. Inorg. Chem.* **2003**, 2153–2163.

[34] U. Vogel, P. Hoemensch, K.-C. Schwan, A. Y. Timoshkin, M. Scheer, *Chem. Eur. J.* **2003**, *9*, 515–519.
[35] U. Vogel, A. Y. Timoshkin, M. Scheer, *Angew. Chem. Int. Ed.* **2001**, *40*, 4409–4412.
[36] K.-C. Schwan, A. Y. Timoshkin, M. Zabel, M. Scheer, *Chem. Eur. J.* **2006**, *12*, 4900–4908.
[37] A. Adolf, *Disseration*, Regensburg, **2007**.
[38] K. C. Schwan, *Dissertation*, Regensburg, **2006**.
[39] S. Schulz, *Adv. Organomet. Chem.* **2003**, *49*, 225–317.
[40] B. Neumüller, E. Iravani, *Coord. Chem. Rev.* **2004**, *248*, 817–834.
[41] A. Y. Timoshkin, *Coord. Chem. Rev.* **2005**, *249*, 2094–2131.
[42] T. J. Clark, K. Lee, I. Manners, *Chem. Eur. J.* **2006**, *12*, 8634–8648.
[43] A. H. Cowley, R. A. Jones, M. A. Mardones, J. L. Atwood, S. G. Bott, *Angew. Chem. Int. Ed.* **1990**, *29*, 1409–1410.
[44] C. von Hanisch, F. Weigend, *Z. Anorg. Allg. Chem.* **2002**, *628*, 389–393.
[45] U. Vogel, *Dissertation*, Karlsruhe, **2001**.
[46] M. Bodensteiner, *Diplomarbeit*, Regensburg, **2007**.
[47] W. Massa, *Kristallstrukturanalyse*, 6th ed., Vieweg+Teubner, Wiesbaden, **2009**.
[48] J. E. Daniels, D. Pontoni, R. P. Hoo, V. Honkimäki, *J. Synchrotron Rad.* **2010**, *17*, 473–478.
[49] P. Debye, P. Scherrer, *Nachr. Ges. Wiss. Göttingen* **1916**, 1–26.
[50] H. M. Rietveld, *J. Appl. Cryst.* **1969**, *2*, 65–71.
[51] U. Vogel, K. C. Schwan, M. Scheer, *Eur. J. Inorg. Chem.* **2004**, 2062–2065.
[52] DFT calculations were performed by using the standard Gaussian 03 program suite (M. J. Frisch, et al. Gaussian 03 (Revision D.01): Gaussian, Inc., Wallingford CT, 2004.) B3LYP functional (A. D. Becke, J. Chem. Phys. 1993, 98, 5648-5652; C. Lee, W. Yang, R. G. Parr, Phys. Rev. B. 1988, 37, 785-793) was used together with standard 6-31G* basis set. Effective core potential basis set of Hay and Wadt (P. J. Hay, W. R. Wadt J. Chem. Phys. 1985, 82, 299-310) was used for W atoms. All structures are fully optimized and correspond to the minima on their respective potential energy surfaces. Full theoretical data is provided at the attached CD.
[53] C. von Hanisch, *Z. Allg. Anorg. Chem.* **2003**, *629*, 1496–1500.
[54] M. Bodensteiner, U. Vogel, A. Y. Timoshkin, M. Scheer, *Angew. Chem. Int. Ed.* **2009**, *48*, 4629–4633.
[55] M. Driess, S. Kuntz, C. Monsé, K. Merz, *Chem.-Eur. J.* **2000**, *6*, 4343–4347.
[56] A. Y. Timoshkin, H. F. Bettinger, H. F. Schaefer, *J. Am. Chem. Soc.* **1997**, *119*, 5668–5678.
[57] A. Y. Timoshkin, G. Frenking, *Inorg. Chem.* **2003**, *42*, 60–69.
[58] F. Thomas, S. Schulz, M. Nieger, *Eur. J. Inorg. Chem.* **2001**, 161–166.
[59] R. J. Wehmschulte, P. P. Power, *J. Am. Chem. Soc.* **1996**, *118*, 791–797.
[60] J. B. Hill, S. J. Eng, W. T. Pennington, G. H. Robinson, *J. Organomet. Chem.* **1993**, *445*, 11–18.
[61] D. A. Atwood, L. Contreras, A. H. Cowley, R. A. Jones, M. A. Mardones, *Organometallics* **1993**, *12*, 17–18.
[62] E. Hey-Hawkins, M. F. Lappert, J. L. Atwood, S. G. Bott, *J. Chem. Soc. Dalton Trans.* **1991**, 939–948.
[63] CSD version 5.32, **Nov. 2010**, The Cambridge Crystallographic Data Centre, 12 Union Road, Cambridge, CB2 1EZ, UK.
[64] J. F. Janik, R. L. Wells, P. S. White, *Inorg. Chem.* **1998**, *37*, 3561–3566.
[65] H. V. R. Dias, A. J. A. III, R. L. Harlow, M. Kline, *J. Am. Chem. Soc.* **1992**, *114*, 5530–5534.

8. LITERATURE

[66] N. Kuhn, G. Henkel, T. Kratz, J. Kreutzberg, R. Boese, A. H. Maulitz, *Chem. Ber.* **1993**, *126*, 2041–2045.
[67] A. Stasch, S. Singh, Herbert W. Roesky, M. Noltemeyer, H.-G. Schmidt, *Eur. J. Inorg. Chem.* **2004**, *2004*, 4052–4055.
[68] F. H. Allen, *Acta Cryst.* **2002**, *B58*, 380–388.
[69] M. S. Lube, R. L. Wells, P. S. White, *J. Chem. Soc., Dalton Trans.* **1997**, 285–286.
[70] S. A. Jasper, S. Roach, J. N. Stipp, J. C. Huffman, L. J. Todd, *Inorg. Chem.* **1993**, *32*, 3072–3080.
[71] Oxford Diffraction Ltd., **2006-2010**, CrysAlisPro (different versions), Oxford, GB.
[72] W. A. Paciorek, M. Bonin, *J. Appl. Crystallogr.* **1992**, *25*, 632–637.
[73] R. C. Clark, J. S. Reid, *Acta Cryst.* **1995**, *A51*, 887–897.
[74] Bruker AXS Inc., **2003**, SHELXTL Version 6.22, Madison, WI, USA.
[75] R. Srinivasan, S. Parthasarathy, *Some Statistical Applications in X-Ray Crystallography*, 1st ed., Pergamon Press GmbH, Frankfurt a. M., **1976**.
[76] G. Oszlanyi, A. Suto, *Acta Cryst.* **2004**, *A60*, 134–141.
[77] L. J. Farrugia, *J. Appl. Crystallogr.* **1999**, *32*, 837–838.
[78] G. M. Sheldrick, *Acta Cryst.* **2008**, *A64*, 112–122.
[79] A. Altomare, G. Cascarano, C. Giacovazzo, A. Guagliardi, *J. Appl. Cryst.* **1993**, *26*, 343–350.
[80] L. Palatinus, G. Chapuis, *J. Appl. Crystallogr.* **2007**, *40*, 786–790.
[81] S. van Smaalen, L. Palatinus, M. Schneider, *Acta Cryst.* **2003**, *A59*, 459–469.
[82] L. J. Farrugia, *J. Appl. Cryst.* **1997**, *30*, 565.
[83] E. Keller, **1997**, SCHAKAL99, Freiburg.
[84] K. Brandenburg, H. Putz, **2005**, DIAMOND 3, Crystal Impact GbR, Postfach 1251, D-53002 Bonn.
[85] V. A. Blatov, A. P. Shevchenko, **2010**, TOPOS 4.0 Professional, Samara, Russia.
[86] P. Müller, R. Herbst-Irmer, A. L. Spek, T. R. Schneider, M. R. Sawaya, *Crystal Structure Refinement: A Crystallographer's Guide to SHELXL*, Oxford University Press, Oxford, UK, **2006**.
[87] R. Herbst-Irmer, G. M. Sheldrick, *Acta Cryst.* **1998**, *B54*, 443–449.
[88] A. L. Spek, *J. Appl. Cryst.* **2003**, *36*, 7–13.
[89] H. D. Flack, *Acta Cryst.* **1983**, *A39*, 876–881.
[90] D. N. Dybtsev, M. P. Yutkin, E. V. Peresypkina, A. V. Virovets, C. Serre, G. Ferey, V. P. Fedin, *Inorg. Chem.* **2007**, *46*, 6843–6845.
[91] H. D. Flack, G. Bernardinelli, D. A. Clemente, A. Linden, A. L. Spek, *Acta Cryst.* **2006**, *B62*, 695–701.
[92] T. Hahn, *International Tables of Crystallography, Volume A: Space Group Symmetry*, 4. ed., Kluwer Academic Publishers, Dordrecht, NL, **1995**.
[93] P. van der Sluis, A. L. Spek, *Acta Cryst.* **1990**, *A46*, 194–201.
[94] M. Bodensteiner, M. Dušek, M. Pronold, M. Scheer, J. Wachter, M. Zabel, *Eur. J. Inorg. Chem.* **2010**, 5298–5303.
[95] S. van Smaalen, *Z. Kristallogr.* **2004**, *219*, 681–691.
[96] V. Petricek, M. Dusek, L. Palatinus, *Jana2006. The crystallographic computing system.*, Institute of Physics, Praha, Czech Republic, **2006**.
[97] F. Boucher, M. Evain, V. Petricek, *Acta Cryst.* **1996**, *B52*, 100–109.
[98] U. Vogel, M. Scheer, *Z. Anorg. Allg. Chem.* **2001**, *627*, 1593–1598.
[99] K. Ruff, *Inorg. Synth.* **1967**, *9*, 30–97.
[100] J. Chatt, L. M. Venanzi, *J. Chem. Soc.* **1957**, 4735–4741.
[101] F. Uhlig, S. Gremler, M. Dargatz, M. Scheer, E. Hermann, *Z. Anorg. Allg. Chem.* **1991**, *606*, 105–108.
[102] G. Becker, M. Rössler, W. Uhl, *Z. Anorg. Allg. Chem.* **1981**, *473*, 7–19.

[103] M. Baudler, G. Hofmann, M. Hallab, *Z. Anorg. Allg. Chem.* **1980**, *466*, 71–75.
[104] N. Kuhn, T. Kratz, *Synthesis* **1993**, 561–562.
[105] G. E. Ryschkewitsch, J. W. Wiggins, *Inorg. Synth.* **1970**, *12*, 116–126.
[106] G. Becker, H. Freudenblum, O. Mundt, M. Reti, M. Sachs, in *Herrmann/Brauer Synthetic Methods of Organometallic and Inorganic Chemistry (Phosphorus, Arsenic, Antimony and Bismuth), Vol. 3*, Thieme-Verlag, Stuttgart, **1996**.
[107] International Union of Crystallography, http://checkcif.iucr.org.

Danksagung

An dieser Stelle möchte ich allen danken, die einen Beitrag zur Entstehung dieser Arbeit geleistet haben:

Allen voran Prof. Dr. Manfred Scheer für das interessante Thema der Phosphanylalane und darüber hinaus für die Möglichkeit, die spannende Methode der Kristallographie zu erlernen und anzuwenden.

Prof. Dr. Alexey Y. Timoshkin für die Durchführung der quantenchemischen Rechnungen.

Dr. Manfred Zabel für die Geduld bei zahlreichen kristallographischen Diskussionen; ihm und Sabine Stempfhuber für das entgegengebrachte Vertrauen im Umgang mit den Diffraktometern. Dres. Eugenia V. Peresypkina und Alexander V. Virovets für alle Ratschläge und das Korrekturlesen des Kristallographieteils.

Dr. Gábor Balázs für die tiefgehenden und interessanten Diskussionen.

Dr. Michal Dušek, Prof. Dr. Václav Petříček und Dr. Lukáš Palatinus für die Einblicke in die Welt der modulierten Strukturen.

Dr. Richard Layfield für die Englischkorrekturen.

Allen Mitarbeitern und Kooperationspartnern des AK Scheer für die Zurverfügungstellung von Kristallen, deren Strukturlösungen Teil dieser Arbeit sind, für die gute Zusammenarbeit und das angenehme Arbeitsklima.

Meinen Praktikanten für die präparativen Beiträge.

Den analytischen und technischen Abteilungen der Universität Regensburg, insbesondere Fritz Kastner und Georgine Stühler für die Geduld bei langwierigen NMR-Untersuchungen sowie Wolfgang Söllner für die Massenspektren bei verminderter Ionisierungsenergie

Meiner Familie für die stete Unterstützung auch in schwierigen Zeiten.

i want morebooks!

Buy your books fast and straightforward online - at one of world's fastest growing online book stores! Environmentally sound due to Print-on-Demand technologies.

Buy your books online at
www.get-morebooks.com

Kaufen Sie Ihre Bücher schnell und unkompliziert online – auf einer der am schnellsten wachsenden Buchhandelsplattformen weltweit! Dank Print-On-Demand umwelt- und ressourcenschonend produziert.

Bücher schneller online kaufen
www.morebooks.de

VDM Verlagsservicegesellschaft mbH
Heinrich-Böcking-Str. 6-8　　Telefon: +49 681 3720 174　　info@vdm-vsg.de
D - 66121 Saarbrücken　　　Telefax: +49 681 3720 1749　　www.vdm-vsg.de

Printed by Books on Demand GmbH, Norderstedt / Germany